Dirt Track Racings Best Kept Secrets

What the Pros Know But Won't Tell the Average Racer!

By Kevin Katzenberg

Dirt Track Racings Best Kept Secrets
Copyright © 2013 by Kevin Katzenberg

All rights reserved. No part of this book may be reproduced or transmitted in any form or by any means without written permission from the author.

ISBN -13 (978-0615919737)

Table of Contents

Preface 5
Introduction 8
Chapter 1 (The Big Three Factors)........................11
Chapter 2 (Wheel Alignment)............................... 13
Chapter 3 (The Panhard Bar and Wheel Alignment) 19
Chapter 4 (Weight Transfer)................................... 22
Chapter 5 (Rear Weight Transfer)........................... 28
Chapter 6 (Traction).. 45
Chapter 7 (Slip Angle)... 46
Chapter 8 (Slip Ratio).. 49
Chapter 9 (Traction Circle)...................................... 50
Chapter 10 (Front End Traction)............................. 53
Chapter 11 (Camber)... 55
Chapter 12 (Camber Gain)...................................... 57
Chapter 13 (Caster)... 59
Chapter 14 (Caster Gain).. 60
Chapter 15 (Toe)... 61
Chapter 16 (Bump Steer).. 63
Chapter 17 (Ackerman Steering)............................ 68
Chapter 18 (Weight Balance and Distribution)....... 70
Chapter 19 (Left Side Percentage)..........................72
Chapter 20 (Rear Weight Percentage)................... 74
Chapter 21 (Wedge)..77
Chapter 22 (Right Rear De-Wedging)................... 82

Chapter 23 (Soft Right Front Spring)................... 84
Chapter 24 (Track Example).............................87
Chapter 25 (Some Swing Arm Notes)...................91
Chapter 26 (Conclusion)................................104

Preface

This book is a culmination of twenty plus years of blood, sweat and tears while in the trenches working on race cars. I've actually been around racing my entire life, but the first half I spent as a grandstand dweller, always looking for a way to get into the action. I began working on cars in 1987 while still in high school. That's usually how it starts, going in the pits as a fan to help someone scrape mud or change tires. Then, there is usually a split that takes place. Some grow weary of the huge amount of work that is involved in racing and head back to the stands. Others get bitten by the bug or hooked on the racing drug and are pulled in deeper and deeper.

For me it was hooked by the drug. I always had an interest in all things technical. I just never got into working on cars. The drug for me was learning about making cars handle better, go faster, and win more races.

Even as I sat in the stands as a kid I could pick up on chassis attitude and could often pick winners against my dads friends who where keenly focused on drinking beer and betting on who would win. It was a good feeling at twelve years old going home with a couple extra bucks when you managed to pick a winner from the last row.

That was a time of qualifying, full field inverts, and weak drunk driving laws. My dads friends would go home with a nice buzz and little lighter in the wallet. But let's get back on track.

Many times the attitude of the car is a dead giveaway to how fast the car is going to be. There is so much to be said about the attitude of the car Many times you can just tell when a car is going to be fast just by how it looks in the first half of a lap.

After an initial introduction to the very basics on the car I tore into everything I could find to read and listened keenly to advise from some of the fastest guys at the track. I learned to separate when people were pulling your leg from when they had something actually useful to say. I figured out pretty early on the core secret that will make your car fast and get it back if you ever get off base. I'm going to give it to you here in three simple words. It is what drove me to write this book and it is what drives the best chassis gurus and engineers in the business. It was mentioned to me by a former Formula one engineer as the most important aspect in racing.

Always Know Why!

The Formula One engineer was a guy named Claude Rouelle. He held an intensive chassis dynamics class I attended in 2005, I think. At the beginning of the class he made a statement that I took as my motto and guiding light to keep me plugging away and learning.

" It's better to finish second and know why than win and not know how or why."

If you know why you can always make improvements to win. If you don't know why you may fall and never figure out how to get up and win.

At First I spent years working on cars as a hobby. I studied everything from sprint cars to modifieds and eventually late models. I even had a brief stint at driving until I quickly ran out of money and realized it really wasn't for me anyway.

In the fall of 2002 I got a call from a friend of mine who was starting a car building business with a partner and they were looking for a welder.

I jumped at the opportunity. Think of it, getting paid to build race cars. What a dream come true. Well the partnership broke up and my friend ended up leaving the company. I took over running the race car shop at Wild Incorporated.

In 2008, when former World 100 winner Dan Schlieper started running our cars, I decided to begin another project. I started a blog to get some of my ideas out in the public in an attempt to help up and coming racers speed up their learning curves.

Many of the topics I brought up in my blog were things I could see happening with cars that no one else was talking about. Things I think the pros had in the back of their mind, but never put down in words for the rest of the racing community to use and help build their racing programs. Things that not only made their cars better, but ideas that made them better as a team.

This book focuses on those principles only pertaining to the car. Maybe there will be future books on different topics, but for now this is the only book I plan on creating.

Well, moving forward, I will continue to explore ideas and blog about them from time to time. I'm sure someday I will come up with thoughts that will rebut what is written here. For now this is what I have to offer the racing world.

Introduction

Learn the secrets that will take you from being in average racer to running at the front every night.

If you have…

- Struggled keeping your car consistently fast every night.

- Wondered what the front runners know that makes some go sailing past you every night.

- Struggle understanding how all these changes will affect the handling of your car.

- Made every effort to keep your car fast every night, but still came up short.

- Listen to popular advice in the pits only to walk away scratching your head.

Then this book holds the solution for you.

Over the past 20 years of working on race cars I have come up with a system of thinking that will get you on the path to winning. Here are some core take aways you can expect to get from this little book.

- Learn the core fundamentals to what makes the cars consistently fast. I have broken it down into a system of three core factors and build everything on to of those three.

- Separate the 'old wives tales' from actual set up changes which will help your car and know why they will help your car. There is far too much information out there surrounding race car setup which is just plain misleading I'm going to cut through all that and give you the basic understanding to making your car work.

- Getting actual system of thinking that will enable you to accurately tune your car to changing track conditions. Part of the problem with race car tuning is the lack of a proper process flow on how to tune your race car.

- Learn about cutting edge technologies and ideas to get up to speed with the rest of the crowd. I don't believe in tricks or secret magic to get a car going fast. It is all just physics, but as we get a better handle on the physics surrounding the suspension, our ideal setup scenario changes.

This is not a book for absolute beginners. There needs to be a basic knowledge of dirt car setups in order to get everything you need out of this book. There are other books on the market which will give much of the basics. Or, just hanging around racing for a couple years will usually give you enough knowledge to fully appreciate what this book has to offer.

There are also some very useful racing seminars. I would highly recommend a live seminar. I've been to several and the learning experience at one of those is absolutely priceless.

If you have been racing for more than a year and still struggle to consistently crack the top five, this book is for you. But, I also urge you use this book as a reference guide. You really don't have to read it from cover to cover to get useful information out of it.

Skip around and pick away at it as you problems arise or you need help in a certain area.

Here is something else to keep in mind. Information and thinking around race car setup is continually changing. If you are reading this book years after the initial publication date, some of the actual tuning methods may have changed and adapted to the way the cars are currently being built. But, the principles are what matters.

Chapter One

The Big Three Factors

In reality there only three things that make a difference in tuning up your race car. All of the things you do , which people tell you that you have to do, falls into those three.

There will be a little overlap between the three area, but if you concentrate on the main three, you won't go wrong.

Ok, now for the big reveal.

Wheel Alignment
Weight Transfer
Traction

The smaller changes you make all relate to these three areas.

And usually it is best to think about these things in a dynamic sense rather than static.

Makes sense right? Your car runs around the track at a completely different attitude than in the shop ... so why limit our thinking to shop thinking only.

Some of the usual changes that fall into these three categories are panhard bar adjustment, bar angles, spring changes, shock changes, caster, camber, toe, tire choices, weight balance, left side weight percentage, rear weight percentage, wedge, fifth coil six coil placement, or pull bar placement.

Don't worry, we will go through all the these changes and describe what categories they affect and how to use them to the tune the car.

Since these usually are what you're dealing with at the race track, this is what most people consider the most important. The main thing you really need to keep in mind is trying to maintain a good balance to the car. Just use these to maintain the balance while you are changing you car to maintain the balance.

But I guarantee if you keep the mindset of the big three areas you will be able to figure out any race car suspension and get anything to handle well.

First, I want to break each one of these down into individual categories and how these effect you cars balance. I'll show you how these relate with each other.

Then we'll build it all back up so you get the overall sense on how to get a good balanced race car.

I should take a quick minute to describe balance. The car, in a very simplified sense should run like you are on the freeway. It should respond to inputs at the wheel without getting loose or tight. It should also want to turn the corner, while applying the throttle, without being loose or tight.

You shouldn't have to snap the gas, stab the brake or flick the wheel to get it to turn. It should be a smooth transition. I mean, in a perfect world, we would want to turn the corner without even lifting the gas.

First we'll start off with Wheel alignment.

Chapter Two

Wheel Alignment

First let's talk about wheel alignment.

If you look at that your car from the back you'll notice that the rear tires do not line up with the front, at least most dirt cars don't. Where these tires point is crucial to the handling of the car.

If you move the left rear tire outside of the left front ... it will create a force to push the left front towards the wall. This will make the car tight or want to understeer.

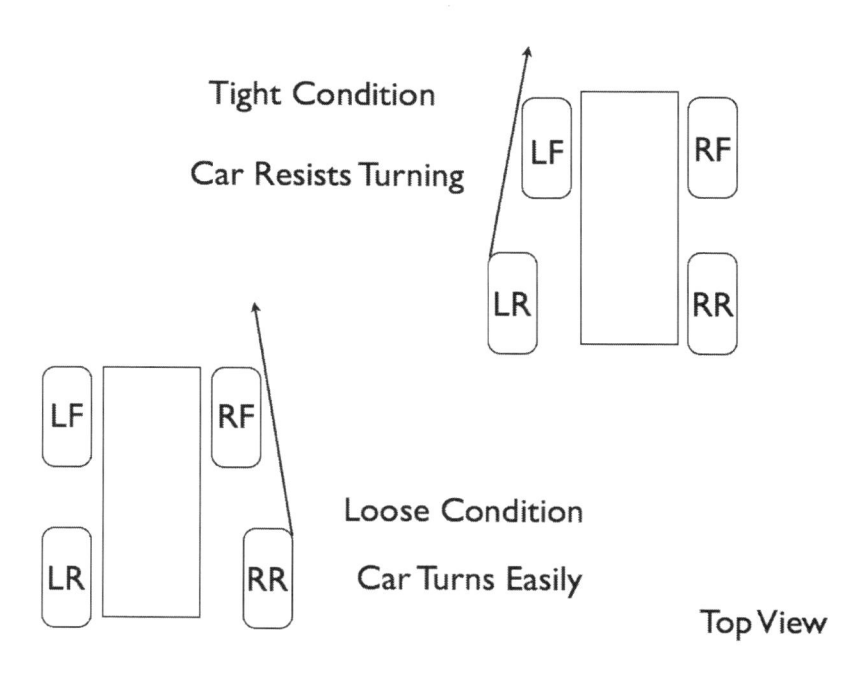

Tight Condition

Car Resists Turning

Loose Condition

Car Turns Easily

Top View

If you push the right rear outside of the right front you will create a loose condition or oversteer.

You can adjust this by ...
- Sliding the rear end in the car left to right.
- Changing different offset wheels in the car.
- Changing axle tube lengths in the car.

I'm dirt it is fairly common to lineup the left side tires or maybe have the left rear talked in from the left front is just a bit on asphalt it is common to have the right side tires lined up but in reality as we get further down the path you will notice how important the right side tire alignment is on dirt.

Actually right side tire alignment is all we should be worried about on dirt or asphalt. With all the role steer in body roll happening in dirt cars now; paying attention to the right side tire alignment is increasingly more important.

Even when the left side tires on the dirt car statically line up in the shop, it really doesn't stay that way when the car is racing on the track. Let's take a look at a floating four link suspension commonly found on modifies and late models.

In the engineering circles this is commonly referred to as a decoupled suspension.

The torque reaction that naturally occurs in the rear end is separated from the locating linkage that hold the rear end from going for and aft. (front to back)

The rear and will then need a separate torque reaction device. This torque absorbing device can either be solid or spring loaded and also have some sort of shock dampening included.

I will go over torque absorption later because it actually falls into the traction category.

So, let's get back to the wheel alignment.

Okay, you're looking at your rear end from the back and let's say the left side tires line up.

Now we begin to look at what happens to the car as it travels around the track.

The driver turns the wheel left, or gives it so wheel input as I like to call it, and the weight transfer begins. We can visually see this happening through body roll. Don't mistake body roll for weight transfer. Body roll is actually just an indicator of weight transfer.

The left side of the car raises in the right side drops. Pretty common sense stuff ... right.

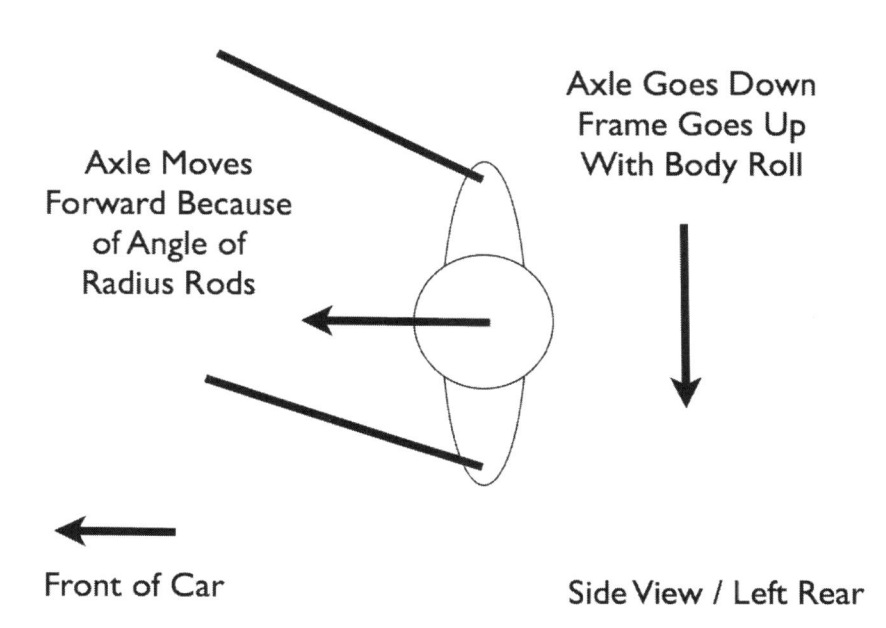

Axle Moves Forward Because of Angle of Radius Rods

Axle Goes Down Frame Goes Up With Body Roll

Front of Car

Side View / Left Rear

Well, on the four link decoupled suspension the left rear tire will go forward and the right rear will go back. This will point the centerline of the rear end toward the outside wall.

If you sight down the left side tires you will actually notice that they no longer line up.

Here is a little side note. I'm going to give you a quick and dirty way to check your wheel alignment. I'll go over the actual procedure to stringing your car all later. But this will be a good way to give a quick spot check at the track or in the garage if you think there's something wrong.

First you want to kneel down behind your car and site down the side wall of the left side tires. Keep moving your head left to right into your lineup the front and back of your rear tire. There will be a point where the front of the tire just disappears. At this point look past the rear tire and see how much the front tire is exposed.
If you can't see the front tire that probably means that the left rear tire is stuck outside of the front. If you see the front tire you can often times judge how much the front tire is stuck out from the left rear. Sometimes you can have somebody help you and put a mark on the front tire then measure it ... but most times you can be pretty accurate just with an educated guess. I also use this method on the right rear on dirt cars ... it is probably easier because 9 times out of 10 the right rear is inside of the right front.

I use this technique all the time especially at the track one helping guys out. Be sure you keep track of whether the rear end is pulled forward or pull back statically to alter the wheelbase side to side. It is only adding to or diminishing what is happening dynamically with the car.

At the end of the day it only really matters what is happening with the role steer in the car dynamically.

Okay... Let's get back to our four link suspension example.

The rear end is pointed towards the wall. So, if you sight down the left side tires now you will notice that the left rear is actually pointing inside the left front tire ... And of the right rear is pointing outside the right front tire.

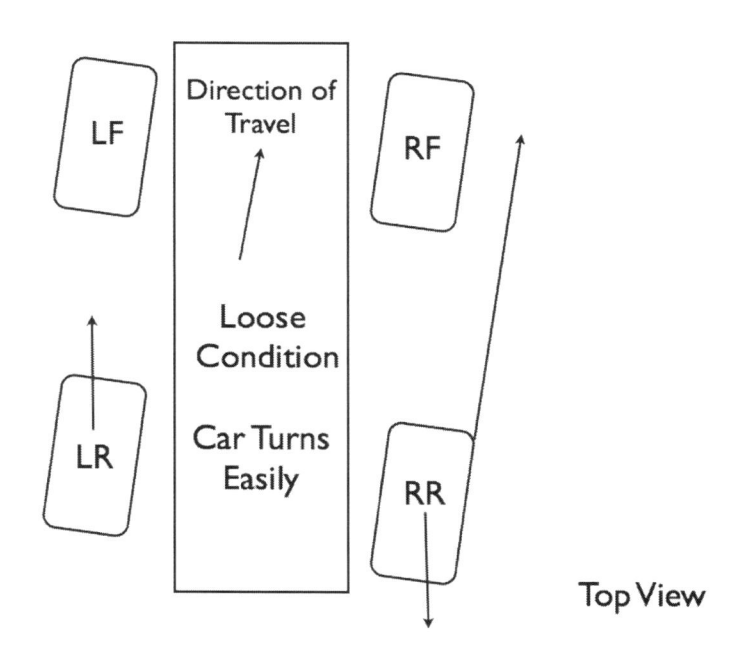

```
         LF        Direction of        RF
                     Travel
                        ↑

                      Loose
                    Condition

                    Car Turns
         LR          Easily           RR
                                              Top View
```

Remember in the previous section where if the right rear is outside of the right front it will cause an oversteer or loose drive condition.

This is not entirely bad. If the car is balanced in other aspects of the set up ... this can provide a good driving car depending on the configuration of the track.

There are many factors that can effect how much 'steer' is put in the rear.

- The height of the radius rods off the centerline of the rear end.
- The rod angles leading from the rear end to the frame.
- The radius rod lengths.
- The split on the radius rod attaching points on the frame.

Here are some general rules of thumb about the adjustment of the radius rods for roll steer.

- The longer the rods … the less role steer the car will have.
- The more angle in the left rear lower rod … the more roll steer is in the left rear.
- The more uphill angle that is in the left rear rods the more loose roll steer the car will have.

Chapter Three

The Panhard Bar and Wheel Alignment

Now that we have that part of roll steer down let me throw a little twist into the wheel alignment section. The panhard can complicate things sometimes. Most times on a four link dirt car this is called a j-bar.

This attaches to the frame on the left side (somewhere near the left side radius rods) and on the rear end just to the right of the driveshaft yolk. A common starting point is at the yolk centerline height off the ground and the frame height is commonly 5 to 10 inches higher than the yolk height. We normally call this the panhard bar split.

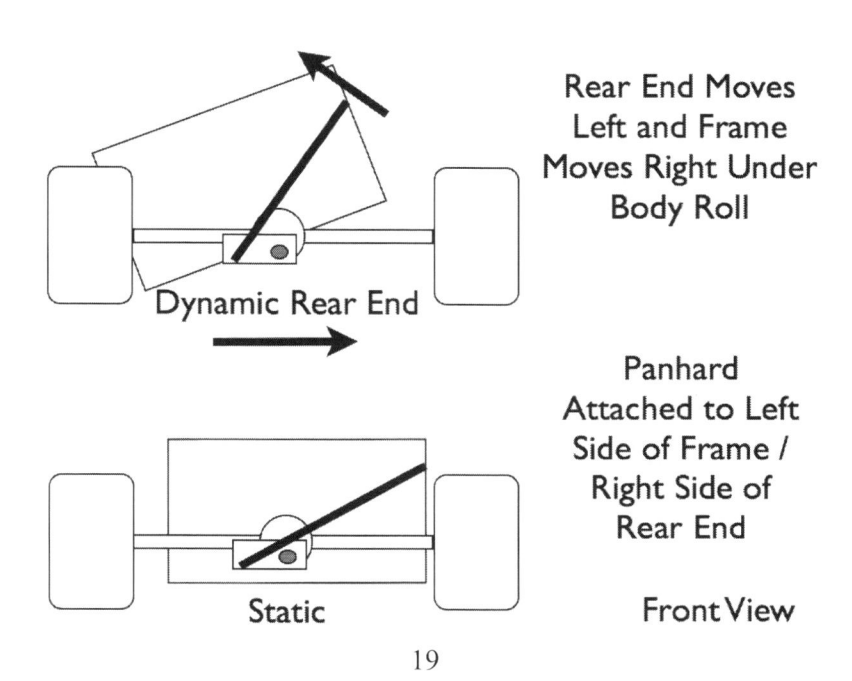

Dynamic Rear End

Static

Rear End Moves Left and Frame Moves Right Under Body Roll

Panhard Attached to Left Side of Frame / Right Side of Rear End

Front View

As the car rolls with the J bar in the car the rear end will

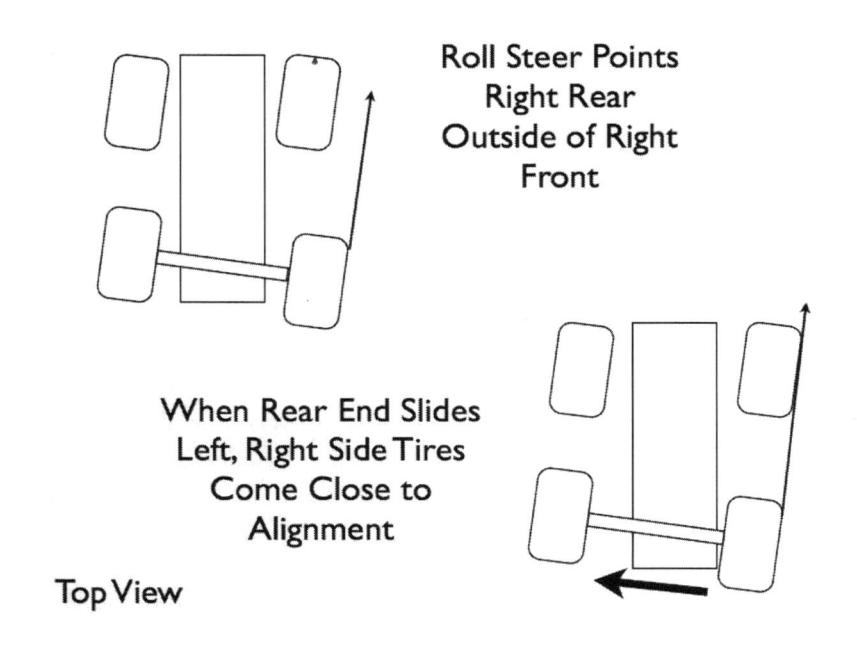

Roll Steer Points
Right Rear
Outside of Right
Front

When Rear End Slides
Left, Right Side Tires
Come Close to
Alignment

Top View

want to slide to the left.

    Now if you look at your car from the back ... the right rear tire that was once outside of the right front is now tucked inside of the right front tire. This will take out that loose roll steer situation and tighten the car back up.

    Here are some factors that will affect wheel alignment with panhard bar adjustment.

- The higher the Panhard bar on the frame... the more the rear end will be pulled left.
- The more overall angle the panhard bar has ... the more the rear end will pull left.
- The shorter the panhard bar ... the more the rear end will move left.

Remember through all these adjustments the key point we want to keep track of is how the right side tires line up dynamically.

All of these adjustments to radius rods and Panhard bar also have other affects on the car. I'll go over that later in the book.

Let's move into the wonderful world of our second most important factor ... weight transfer.

Chapter Four

Weight Transfer

Let's really dig into weight transfer. First, as I stated earlier, body roll is not weight transfer it's only the indication of weight transfer.

On a go cart with no suspension … weight still transfers whether or not there is any body roll. Because go carts have no suspension, they rely heavily on chassis flex to tune. If you watch a go cart from the rear you will notice they often lift the inside tire off the track to prevent scrub and allow the cart to turn. They use weight transfer and chassis flex to pick up that inside tire. Without weight transfer this would never happen.

What Effects Weight Transfer?

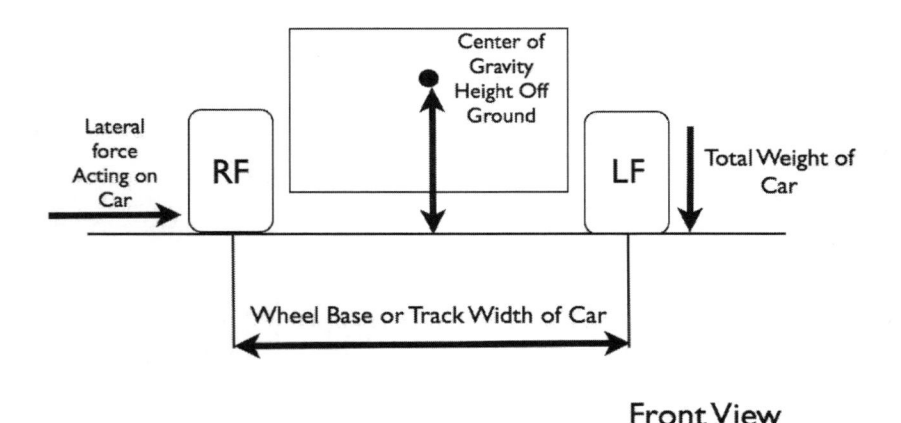

Front View

Weight transfer is only affected by four things in a car.

- The center of gravity height in relationship to the ground.
- The amount of "G" forces acting on the center of gravity.
- The weight of the car.
- The wheelbase or track width.

Look at it this way.

If we alter any of these three things we will alter the amount of weight transfer in the car.

For instance ... If we stack lead high in the car we are effectively raising the center of gravity.

Raising the center of gravity will increase weight transfer. The thing that some people fail to realize is that raising the center of gravity will also slow down the weight transfer in the car. It will also take more side bite and more traction to get that weight moving.

Think of your roll center as the center of a tire. Let's say you have a 6 inch diameter tire next to an 18 inch diameter tire. Place your hand at the top of each tire and try to spin them. There'll be several things that you will notice.

First ... It will take more force for your hand to start the larger tire rotating then the small tire.

Second ... It will be harder to stop the larger tire from spinning.

Third ... It will take more time for the larger tire to accelerate up to speed.

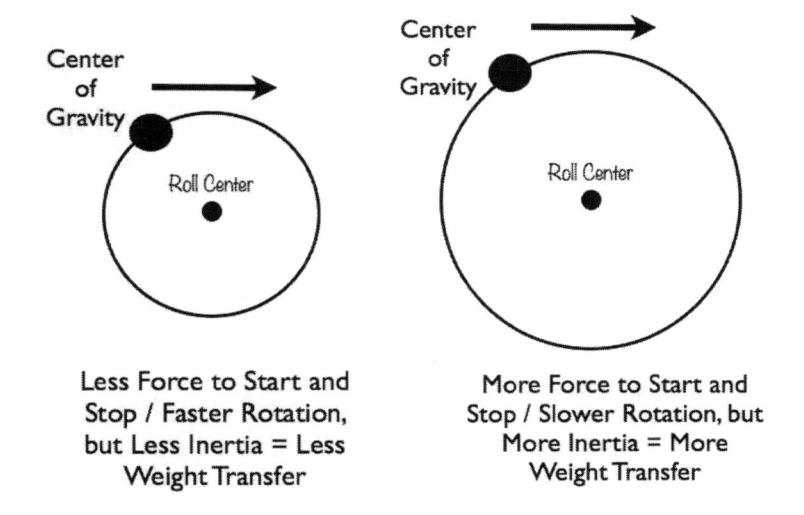

Center of Gravity → Roll Center

Center of Gravity → Roll Center

Less Force to Start and Stop / Faster Rotation, but Less Inertia = Less Weight Transfer

More Force to Start and Stop / Slower Rotation, but More Inertia = More Weight Transfer

Another way to actually see the speed of rotation is what happens in ice skating. As a skater begins to spin. The tighter in they pull their arms to the center of their body, the faster they will rotate.

The further away they move their arms from their body the slower they will rotate.

Now think what would happen if they had a 20 lb. barbell in each hand and they slowly started to extend their arms away from their body.

It would be more difficult for them to start and stop their rotation because the barbells would want to continue their rotation more than if their hands were empty.

Now let's look at how these factors affect our car as it goes around the track.

We noticed it takes more force to get the larger tire rotating. The force produced by your hand on our example can actually represent forward bite and side bite.

So ... In order to get higher wait to transfer and lower weight you need more side bite and more forward bite.

If it takes more force to stop weight transfer ... then as weight is transferring on to the springs it will take more shock dampening to keep the car from bouncing on the springs when weight is moved high.

If it takes a longer time to get higher wait to transfer ... then the car reaction time is slowed down the higher the weight is in the car.

All three of these are true regardless of whether we are taking about weight transferring front to back on acceleration or side to side when we start to turn the corner.

Elastic vs. Kinematic Weight Transfer

Now that we have that established we need to define that there are actually two types of weight transfer.

The first one is what I call kinematic weight transfer. This is the transfer that happens almost immediately through solid suspension locating linkages. This includes weight transfer through things like rear and locating linkages, panhard bars, and upper and lower control arms. I say that this is instant weight transfer although that's not entirely true. It does happen considerably faster than the other type of weight transfer.

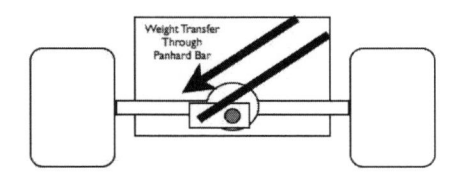

Kinematic Weight Transfers Quickly Through the Linkages

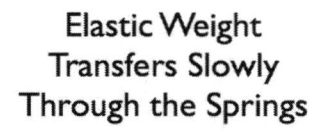

Elastic Weight Transfers Slowly Through the Springs

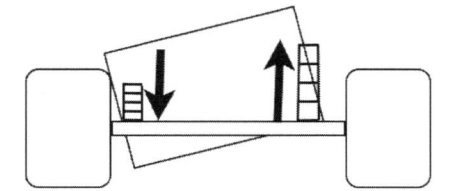

The other type of weight transfer is felt in the car through springs and shocks I called this type elastic weight transfer. As we discussed earlier, body roll is an indicator of weight transfer. It takes time for the car to react and roll onto the springs.

Here is the way I see the relationship between the two.

Kinematic weight transfer is more responsive and can probably be felt almost instantly with the flick of the wheel or stomping of the gas. The downside to this is that there is little to no cushioning to the tire contact patch. The tire itself becomes a cushioning for bumps in the track surface.

Elastic weight transfer is slower to react but provides good cushioning to the tire contact patch. This cushioning is essential to keeping traction in the tire especially on rough racing surfaces.

Just like in the previous section on wheel alignment, there are some core factors we need to keep an eye on for ideal car performance.

We need to have enough body roll to keep the right side springs compressed and providing a cushion to the contact patch. This is becoming increasingly more difficult without going overboard and having the right side cave-in on the springs. What I mean is too much spring compression. If I were forced to throw a number out there as to how much spring compression is necessary … I would have to say between 2 1/2" and 3" on the right rear spring.

Chapter Five

Rear Weight Transfer and Roll Center

Now we will get into one of the most difficult and most often misunderstood aspects of the car. It took me years to put together what is actually going on in the rear of a modern 4-Link race car and I'm expanding my understanding everyday.

With all of it's complexity it is impossible to say something has to be a certain way to get the best results. A small change in panhard bar, spring rate or side bite can change the entire equation. What I really intend for this section is to give you the best information possible so you can make the best decisions possible for your race car.

This section exclusively deals with the common decoupled 4-Link suspension which uses a J-bar mounted on the left side of the car and the right side of the pinion on the rear end.

If you run a swing arm suspension you can get a detailed description how they differ in the swing arm section.

Panhard Bar and Elastic Weight Transfer

One of the most common adjustments we use to adjust weight transfer is the Panhard bar height and angle.

Adjusting J-bar angle can have a huge effect on kinematic tire loading. As a rule of thumb, the more angle on the J-bar the more kinematic transfer. This has to be balanced with the affect of the J-bar pulling the rear end to the left.

The rear panhard bar is a weird animal, it can do some funny things to your car if you really not careful.

This is the way I see a j-bar working in the car for elastic weight transfer. Picture your car from the back. Your center of gravity height in most cars is up roughly around camshaft height of the engine. And in many race cars I've checked it lies front to back and side to side somewhere near the shift levers.

Now picture a rope attached at imaginary center of gravity, around camshaft height, leading out the right side of the car like you're going to have a tug-of-war. Get a bunch of people together and pull on the rope rolling the car to the right.

These Two Points Will Naturally Want to Even Out providing Some Natural Kinematic Roll to Car. The Further the Split, the Larger Roll Kinematics Plays In Elastic Weight Transfer.

Jacking Effect Provided by the Panhard Bar

Gravity Pulling Down Canceling the Jacking Effect of the Panhard Bar

Amount of Weight Transfer and Right Rear Spring Rate Determine Amount of Compression on Right Rear

Now watch what happens to your car in the rear. Picture an imaginary line leading from the center of gravity height to the J-bar mounting position on the frame.

If the J-bar mounting height on the frame is lower than the center of gravity the right side of the frame roll will naturally want to pull down and level of this plane out. The panhard bar provides assistance to the elastic weight transfer onto the spring. Once this point levels out, the panhard bar no longer aids in elastic weight transfer. It only produces a jacking effect to the left rear and kinematic weight transfer to the right rear.

If the J-bar mounting height on the frame is initially level to the center of gravity there will not be any assistance by the J-Bar to the elastic weight transfer. There will only be kinematic weight transfer through the panhard bar and a jacking effect to the left

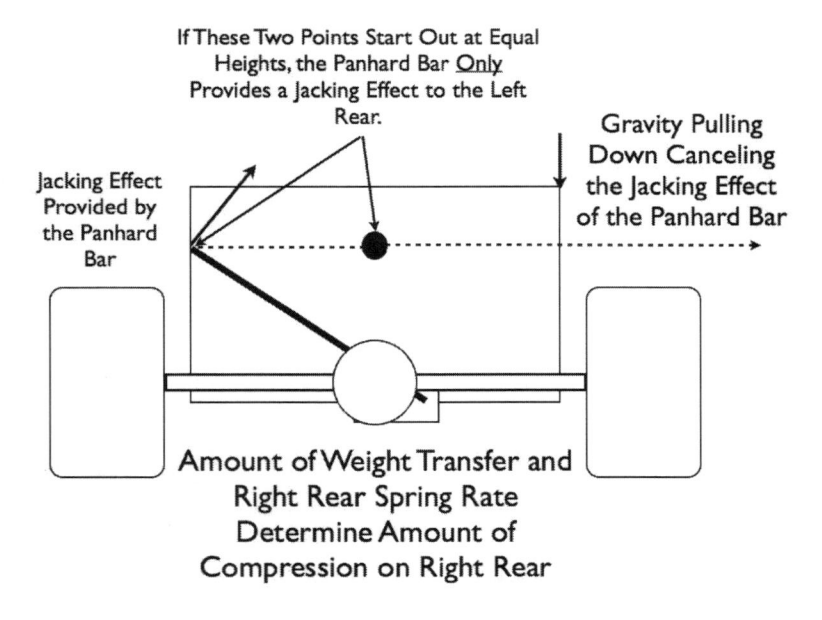

If These Two Points Start Out at Equal Heights, the Panhard Bar Only Provides a Jacking Effect to the Left Rear.

Jacking Effect Provided by the Panhard Bar

Gravity Pulling Down Canceling the Jacking Effect of the Panhard Bar

Amount of Weight Transfer and Right Rear Spring Rate Determine Amount of Compression on Right Rear

rear. The car will not roll as easily as it did when the panhard bar was low.

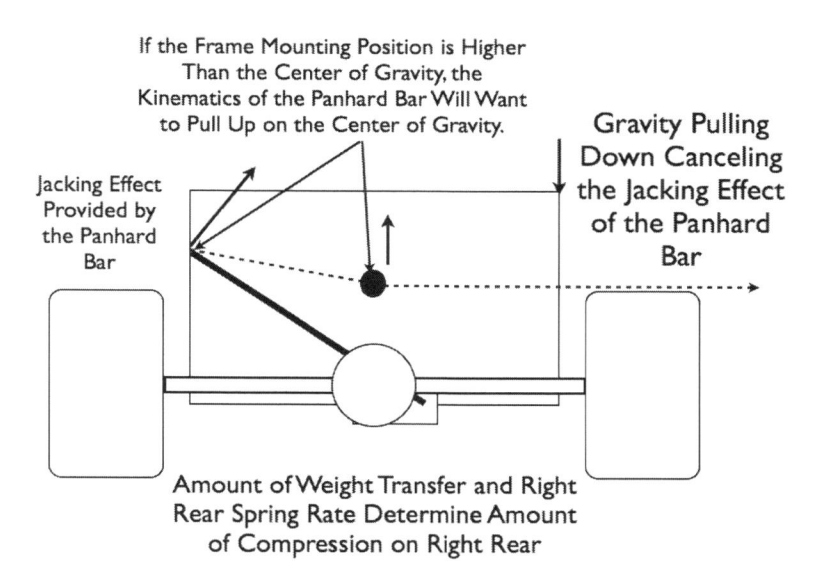

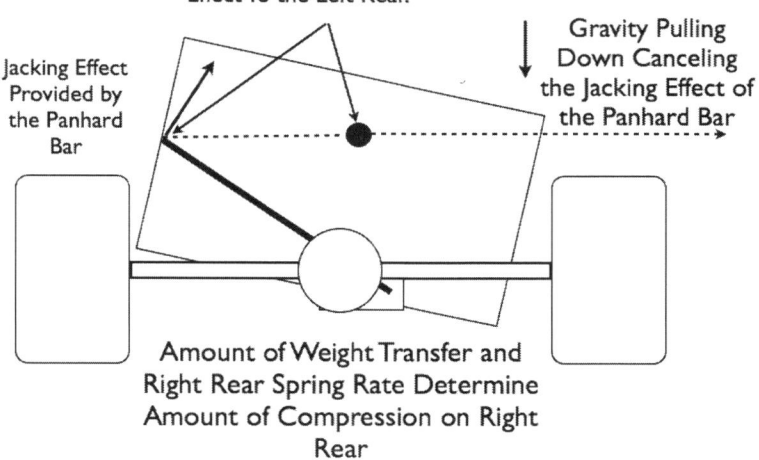

If the Frame Mounting Position is Higher Than the Center of Gravity, the Kinematics of the Panhard Bar Will Want to Pull Up on the Center of Gravity.

Jacking Effect Provided by the Panhard Bar

Gravity Pulling Down Canceling the Jacking Effect of the Panhard Bar

Amount of Weight Transfer and Right Rear Spring Rate Determine Amount of Compression on Right Rear

Once These Two Points Have Leveled Out, the Kinematics of the Panhard Bar Only Supply a Jacking Effect To the Left Rear.

Jacking Effect Provided by the Panhard Bar

Gravity Pulling Down Canceling the Jacking Effect of the Panhard Bar

Amount of Weight Transfer and Right Rear Spring Rate Determine Amount of Compression on Right Rear

Lets say for instance we are working on a car with a low center of gravity or the panhard bar is mounted extremely high on the frame. If the J-bar is mounted higher than the center of gravity height the kinematic forces will actually want to pull up on the right side of the car.

It is unlikely that the car will literally raise up because of gravity pulling down on the right side of the car, but I would guess compression on the right rear spring would be a little less. The car will have a slightly higher resistance to roll because of the panhard bar. There is a force there that limits body roll naturally.

Kinematic Weight Transfer

The More Angle on the Panhard Bar, the Less Resistance to Jacking on the Left Rear

The Less Angle on the Panhard Bar, the More Resistance to Jacking on the Left Rear

The Less Angle on the Panhard Bar, the Less Kinematic Weight Transfer Downforce is Applied to the Rear End.

The More Angle on the Panhard Bar, the More Kinematic Weight Transfer Downforce is Applied to the RearEnd.

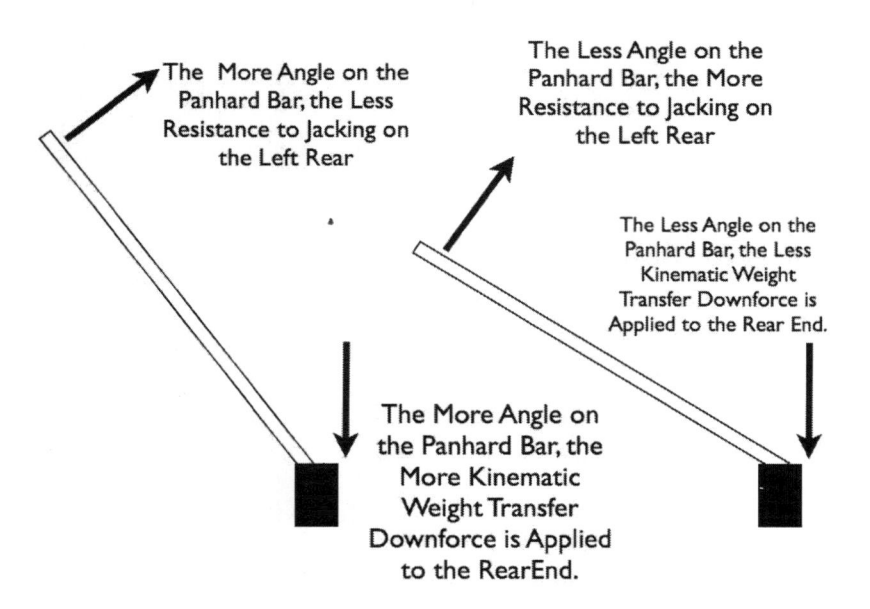

For a J-bar, kinematic weight transfer can easily be illustrated with the use of a common push broom.

The steeper the angle on the handle of the broom the more downforce you put on the bristles. This will cause the bristles of the broom to bend easily. It will also want to lift the handle of the broom easier. In our car this will produce a jacking effect on the left side.

The more shallow the angle of the Broom handle the more the bristles just want to slide across the floor. This will put less kinematic don't force on the bristles. The same is true in our car. Less downforce on the bristles is less kinematic weight transfer but more elastic weight transfer.

Here are some general rules of thumb concerning panhard bar adjustment. These are concerning J-bars. I will discuss other types of Panhard bars later.

- The lower the Panhard bar mounting position on the frame – the more body roll and elastic weight transfer to the springs.
- The higher the panhard bar mounting position on the frame – the more kinematic weight transfer and less natural body roll.
- The more angle on the J-bar – the more kinematic weight transfer.
- The less angle on the J bar – the more it elastic weight transfer through the springs.

Spring Roll Centers

Springs also effect the rear roll center and weight transfer in the rear. I believe spring split will have a greater effect in the front than in the rear, but for sake of a clear understanding on what is going on in the rear of the car, let's talk about them now.

Picture a car with equal springs and they are mounted an equal

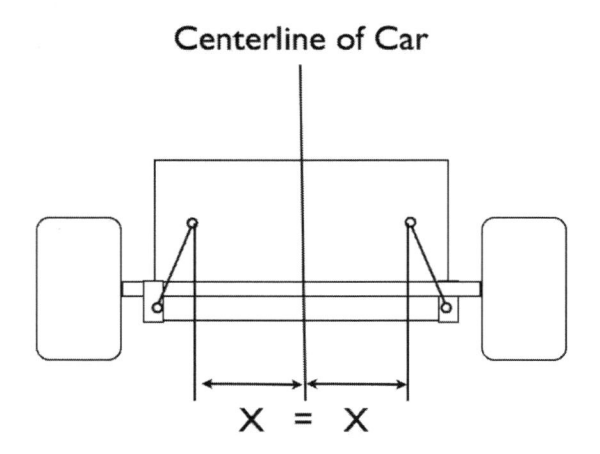

Centerline of Car

X = X

Both Springs are Equal Distance From the Center Line of the Car

distance from the center line of the car and from the rear wheel centerline. This is rarely the case, but for simplicity let's pretend.

Let's also pretend that panhard jacking and everything else we talked about previously doesn't have an effect in our test cases.

As weight transfers from left to right the same amount of weight will want to leave the left rear and compress the right rear. The left rear spring will extend the same amount as the right rear spring compresses. This will put the roll center in the middle of the car.

Centerline of Car

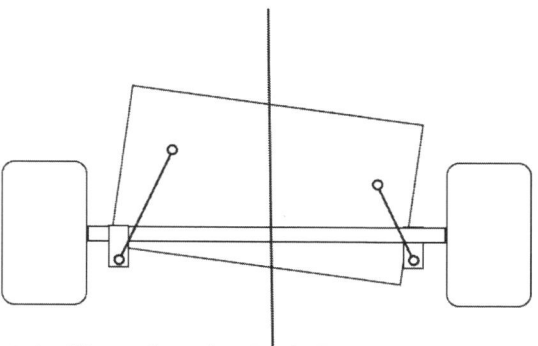

With Weight Transfer the Left Rear Spring is Extended the Same Amount as the Right Rear Spring is Compressed. This Puts the Roll Center in the Middle of the Car.

Now, if we soften the right rear spring and stiffen the left rear spring, given the same amount of weight transfer, the right rear

Centerline of Car

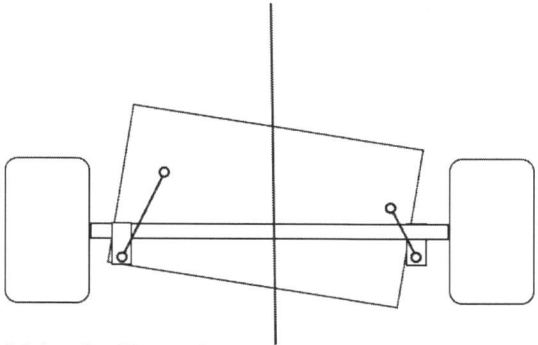

With Weight Transfer the Left Rear Spring is Not Extended as much as the Right Rear Spring is Compressed. This Puts the Roll Center Closer to the Left Rear Spring.

spring will compress more and the left rear spring will extend less. This will move the roll center closer to the left side of the car.

If you have a hard time picturing this, go to extremes. Imagine a 1000 lb. / inch spring in the left rear and a 10 lb. / inch spring in

Centerline of Car

With Weight Transfer the Left Rear Spring is Extended More Than the Right Rear Spring is Compressed. This Puts the Roll Center Closer to the Right Rear Spring.

the right rear. Let us transfer 100 pounds from the left to the right. The right rear will compress 10 inches (if it doesn't coil bind first) and the left rear will extend 1/10th of an inch. Looking at the car from the back, it will appear to have a roll center right at the left rear spring.

The opposite is true when stiffen the right rear spring and soften the left rear spring. The roll center will move towards the right side of the car.

Now keep the same distance between the springs in the car, but shift them to the left so they no longer line up with the center line of the car. The roll center will naturally move left in the car with

Centerline of Car

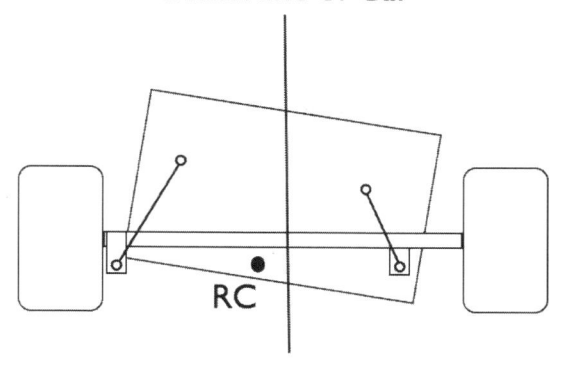

With Weight Transfer the Left Rear Spring is Extended the Same Amount as the Right Rear Spring is Compressed, But Spring Stance is Moved Left. This Puts the Roll Center Left of the Centerline of the Car.

even springs and if we stiffen the left rear spring and soften the right rear spring the roll center will move even further left.

Centerline of Car

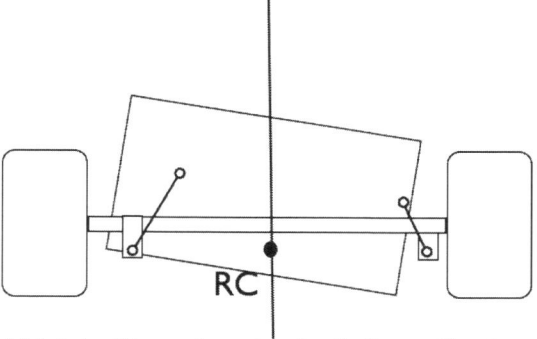

With Weight Transfer the Left Rear Spring is Not Extended the Same Amount as the Right Rear Spring is Compressed, But Spring Stance is Moved Right. This Puts the Roll Center on the Centerline of the Car.

Now it can get really interesting.

Lets move the centerline between the springs to the right, but stiffen the left rear spring and soften the right rear spring. This could put the spring portion of the roll center back on the centerline of the car.

Look at how each example affects the chassis attitude and how the car rolls. This can be a very important tuning tool to adjusting the body attitude during weight transfer while still maintaining certain roll centers.

Trial and error and a little bit of measuring will go a long way to figuring out how the springs fit into the whole equation of the roll center.

Now add up all of the forces mentioned and try to figure out where the rear roll center is on a 'J-bar' panhard bar car. Knowing and understanding what each of these things do when broken down into simple terms will enable you to make the best tuning decisions when faced with handling problems.

Remember of the whole goal of tuning our race car is to have a nice smooth balance that makes it easy to drive fast. Our goal is not to have a car that acts like a carnival ride.

Rear Linkage Weight Transfer

There is a trade off in our rear linkages between loading under acceleration kinematically and loading through the springs with anti-squat.

Thrust Angle of the Rear Suspension

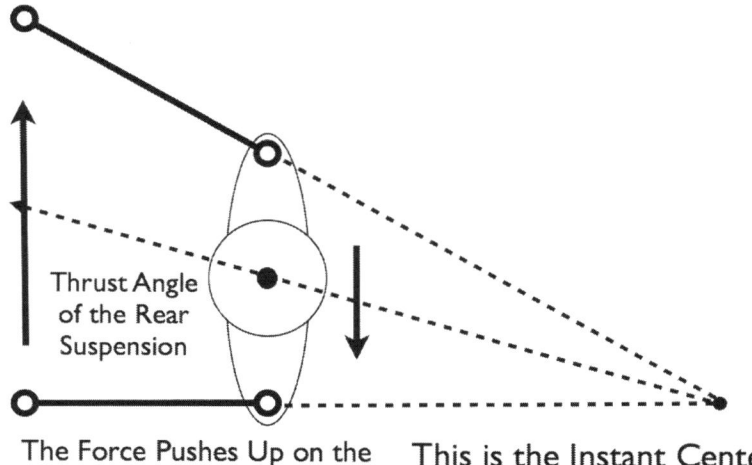

The Force Pushes Up on the Frame and Down on the Rear End; Driving the Tire Into the Track Surface.

This is the Instant Center or "Push Point" of the Rear Trailing Arms.

Let's look at adjusting rear radius rods for kinematic weight transfer or loading on acceleration. There is a good amount of science behind this, but what I'm going to do is give you a general overview and some guidelines and leave you to test and experiment with your car.

In a decoupled 4-link suspension, where the radius rods intersect we will call the push point. The technical term is instant center, but I like push point. Where this push point lies in relation to the centerline of the axle is very important to weight transfer and traction.

Weight transfer from front to back on acceleration will hit the radius rods first. Actually, we need to think about this in terms of the weight hitting the push point or the intersection of the radius rods.

If the push point is below the centerline of the rear axle then there is a force pushing down on the axle as acceleration thrust of

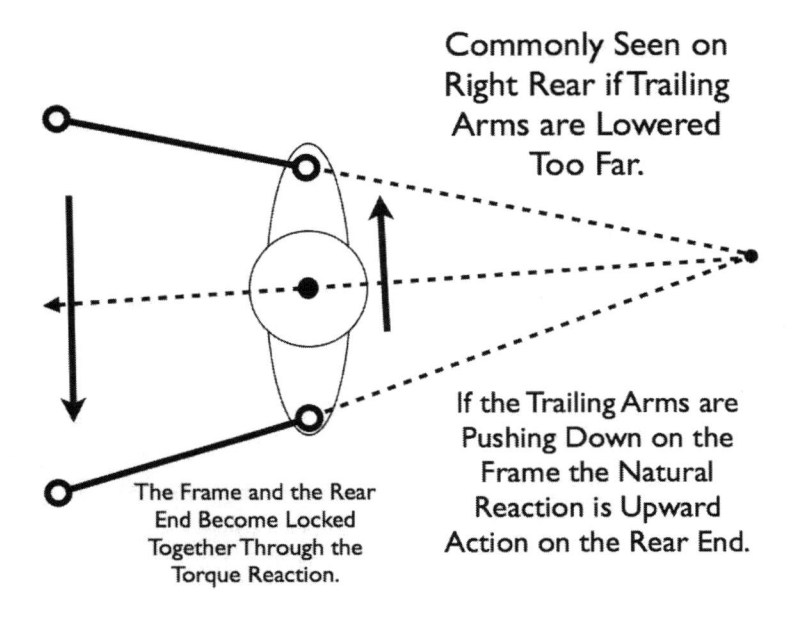

Commonly Seen on Right Rear if Trailing Arms are Lowered Too Far.

The Frame and the Rear End Become Locked Together Through the Torque Reaction.

If the Trailing Arms are Pushing Down on the Frame the Natural Reaction is Upward Action on the Rear End.

the axle pushes up on the frame. Depending on the height of the push point, this can add traction and drive to the rear suspension through weight transfer.

If the push point lies above the centerline of the rear axle it can have a lifting effect. This is often times canceled out by the elastic weight transfer onto the spring. But, there is a little bit of what I call torque or mechanical bind in the suspension system.

As the right rear of the frame is compressed against the spring, the natural reaction is an upward thrust of the axle. Gravity and weight transfer will hold the tire in compliance to the track, but the trust angle of the rear suspension is being used to 'Lock' the axle against the spring. This most often is the case on the right rear of decoupled 4-link race cars.

Ideal Right Rear Dynamic Situation.

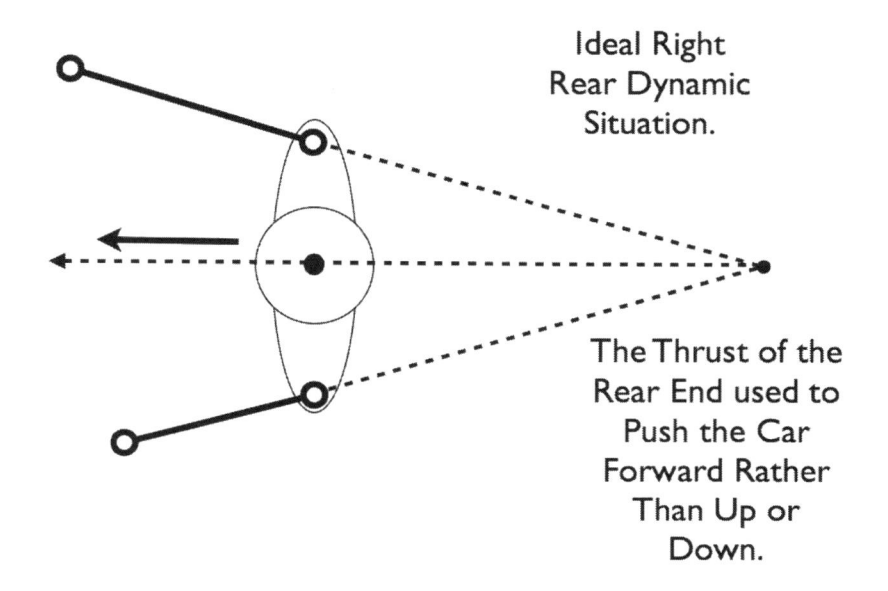

The Thrust of the Rear End used to Push the Car Forward Rather Than Up or Down.

You need to keep enough bar angle out of that corner so the car doesn't lift off the spring with upward thrust. But, enough bar angle is needed in that corner so the car doesn't go into that strange mechanical bind.

Although the car statically will not start out this way, they end up this way all to often through body roll. Balance is critically important.

I believe the ideal race car will run dynamically with the right rear push point on the centerline of the rear axle or maybe just a little bit below.

The natural settling of weight onto the spring, on the left rear, doesn't happen too often anymore. On modern for link cars we run all large amount of anti-squat in the left rear. The left rear radius rods do more for adjustment of roll steer, indexing and kinematic loading of the tire rather than relying on the inertia of the car loading the spring.

Traction on the left rear is often the least of our worries. The biggest problem we have in current times is keeping the right rear spring and the right rear tire loaded. There are some suspension types still running that rely on that inertial spring loading to get traction off the corner. I'll discuss that later in the book.

Front View of a Typical Independent Front Suspension with Upper and Lower Control Arms.

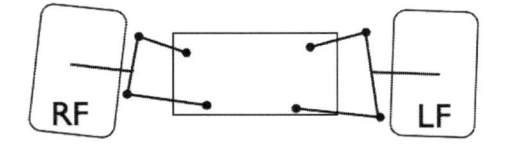

Top View of a Typical Independent Front Suspension with Upper and Lower Control Arms. This is a Strut Style Suspension Typically Found on Late Models.

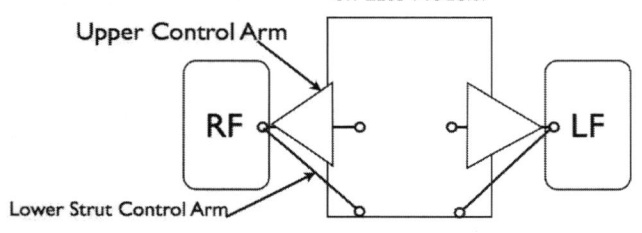

Upper Control Arm

Lower Strut Control Arm

Front Weight Transfer

Front end kinematic's are a little bit different than the rear. The kinematics of the front suspension points lend itself more to elastic weight transfer (transfer through the springs) than weight transfer through the linkages. But, we do still use some kinematic loading to aid traction; especially on the left front tire.

Most dirt car front ends consist of an upper and lower control arm to hold the wheel both laterally and longitudinally. First will go through the effects of how the weight transfer happens laterally.

Front View of a Typical Independent Front Suspension with Upper and Lower Control Arms.

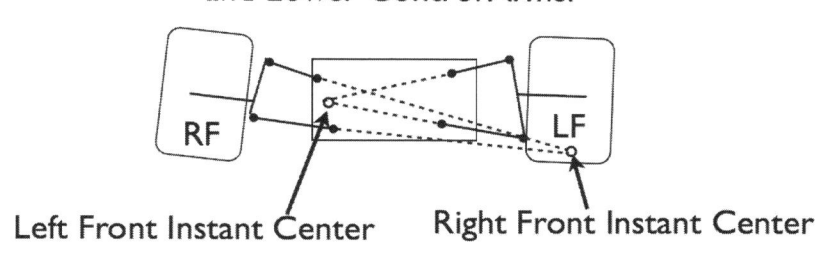

RF   LF

Left Front Instant Center   Right Front Instant Center

When you start to turn the wheel a torque will be applied to the front tires. The torque travels through the front control arms to the attachment points on the frame. Like in the rear, if you extend the control arms, they will meet at an instant center or push point. The height and placement of these will determine the jacking of the front end.

The basic rule of thumb is if the right side intersection point is above the ground it will want to produce a jacking effect on that corner. That's why it is important to get this intersection as close to the ground as possible. The closer this point is to the ground the stiffer right front spring you will need to run because the suspension will feel softer to the car in roll. Cars with a lower right front instant center will need to run sometimes 50, 75 or 100 pounds stiffer with the right front spring.

The opposite holds true for the left front tire. The higher the point is above the ground the less that corner of the car will want to jack under roll. A torque will actually be applied to help hold the left front on the ground. I believe this is the difference between some cars that will lift the left front tire off the ground and some cars it will hey have both front tires glued to the ground.

This anti-jacking effect along with minimal camber change will actually put a kinematic load on the left front tire. I've actually seen firsthand how changing front end geometry can break weak left front suspension components. On cars I've worked on, we've actually had to replace aluminum upper control arm across shafts with steel just for durability.

Chapter Six

Traction

It really wouldn't be right talking about traction without first talking about where traction begins; at the tires.

Tires can get to be a very complicated subject. This book is about basics, practicality and commonsense. So, we will keep our tire discussion practical.

First of all, contrary to what many believe, a tire doesn't always produce more traction the more load you put on it.

But, before we can understand why this is, we need to define a couple of terms.

The rubber in the tire only gets traction in three separate ways.

- Adhesion – the rubber glues itself to the surface of the track.
- Deformation – the softness of the rubber molds around asperities in the track.
- Wear – energy is released when rubber breaks apart and produces friction.

Next we have to understand slip angle.

Chapter Seven

Slip Angle

Slip angle is the deforming of the tire to put a lateral load on the car or side bite in the car.

Imagine putting a tire on a piece of glass and looking at it from the bottom up. Push down on the tire and try to twisted to the left. You'll notice that the contact patch part of the tire will stay stationary and the rest of the tire will twist to the left.

Now begin rolling the tire in you will notice it will turn to the left, but the contact patch will actually be at a different angle. The difference in the angle is called slip angle.

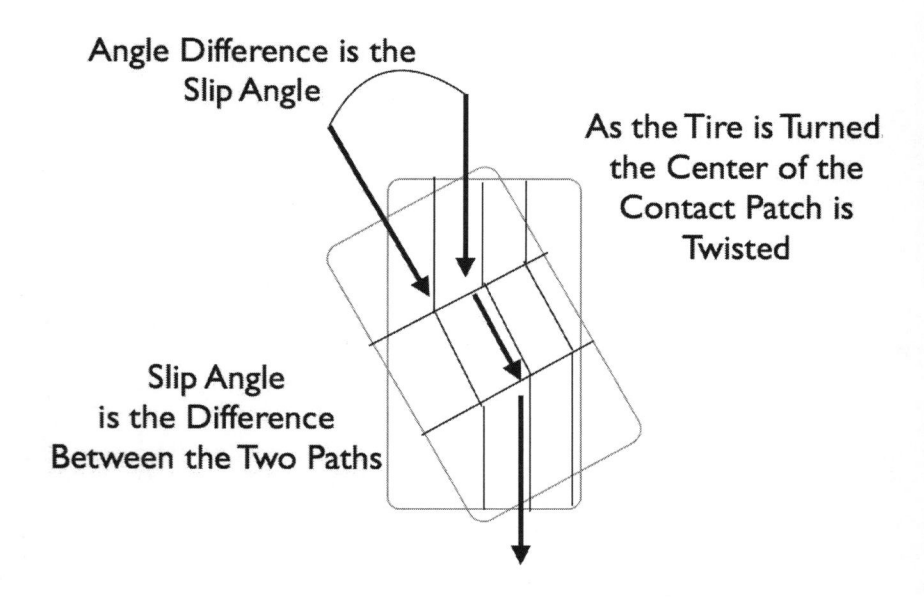

Angle Difference is the Slip Angle

As the Tire is Turned the Center of the Contact Patch is Twisted

Slip Angle is the Difference Between the Two Paths

As the tire rolls… The relaxed part of the tire enters the contact patch and changes it's angle. As the trailing edge of the tire leaves the contact patch it snaps back into its relaxed state. This continual loading and unloading to different angles is what produces traction for the tire to turn.

There is a optimum slip angle which is designed into each type of tire. The most traction that tire will see is when you can run at the highest designed slip angle.

I may be dating myself here he little bit, but if you have ever ridden a big wheel when you were a kid you know what lack of slip angle feels like.

These were three wheeled riding machines where you pedal the front tire. The tires were hard plastic and didn't get much traction or steering at all. If you ever got up any speed with them and turn pretty quick it would just push (or understeer for the more technical crowd).

Now when we grew up a little, our bicycles with air filled rubber tires, turned much better. We also needed to be more precise with our steering. The bicycle would react very quickly to small inputs in steering.

Here are some common rules of foam surrounding slip angle.

- The larger and softer the sidewall of the tire – the more slip angle and more traction the tire would have.
- The more slip angle – the last response of the tire would be.
- The less responsive – the more slushy the car would feel.
- The narrower door wheel with the same tire – the more slip angle it produced.

- Each type of tire created will have an optimum slip angle that it needs to run at for best traction.
- Running at a tires optimum designed slip angle will create the most traction.

There is a trade-off with everything in our race cars. Slip angle just happens to be one.

- Tires with more slip angle – great traction – slushy response.
- Tires with less slip angle – great response – little traction.

There are ways to counteract less slip angle with using other forms of traction with the tire. Running a softer tire compound, siping and grooving patterns are ways we can enhance traction on tires that have a limited slip angle.

Chapter Eight

Slip Ratio

Now let's take a look at slip racial. Slip ratio is similar to slip angle. just in acceleration and braking.

If you have ever seen the slow-motion video of a dragster taking off from the line, you probably have noticed all the rear tires deform under acceleration.

The stretching of the sidewall of the tire will produce a bulge at the leading edge of the tire. The relaxed tread enters the contact patch where it begins to stretch. As the tread leaves the ground at the back of the tire it snaps back into place. This releasing of energy and snapping back is what pushes the car forward.

Let's go back for big wheel example. If you tried to pedal real fast it was easy to spin the front tire because it had no slip ratio with the hard plastic tire. Our bicycle on the other hand could accelerate as hard as we needed it to go.

The same holds true under breaking. As you lift the gas and apply the brakes the tire wrinkles in the reverse direction. The entire process that happens at the front of the tire to push it forward happens at the rear to pull the car back under braking.

Now that we have the basics down ... Let's look at how these relate to actually putting traction in the car.

Chapter Nine

Traction Circle

The whole secret to getting all the traction we need is the load the tire to its maximum traction limit precisely when you need it to provide the traction.

Sounds easy doesn't it ... Well it's not as easy as it sounds because we still need to abide by the laws of physics ... Unfortunately.

I really need to introduce the concept of a traction circle to show you some of the limits you are fighting in this battle for traction.

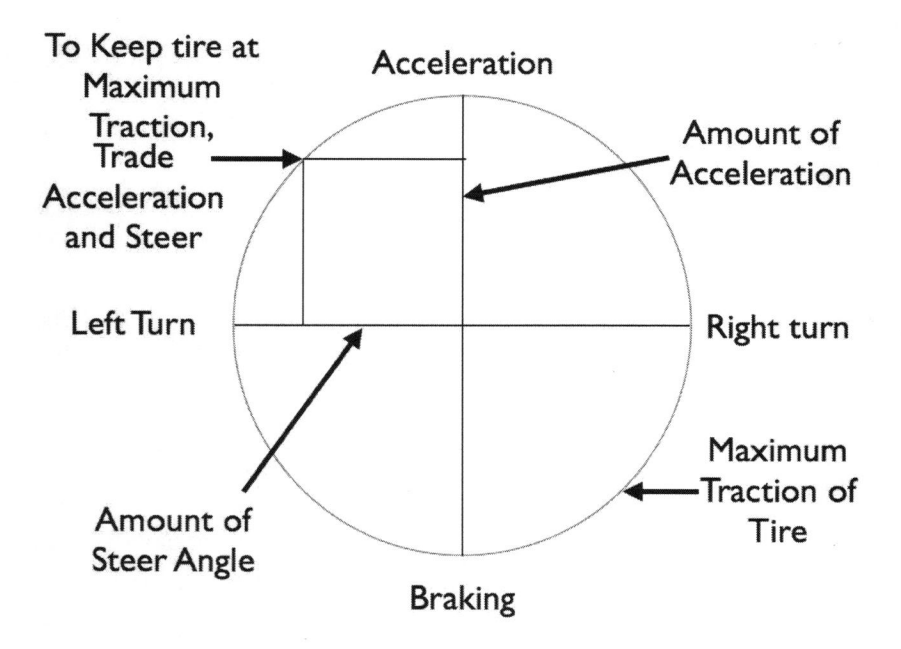

Diagram: Traction Circle

- To Keep tire at Maximum Traction, Trade Acceleration and Steer
- Acceleration
- Amount of Acceleration
- Left Turn
- Right turn
- Amount of Steer Angle
- Maximum Traction of Tire
- Braking

This represents the trade off you need to make when trying to get your car side bite, acceleration traction, and braking traction.

The sidewall of the tiger wants to flex in three different directions to achieve these three things. So, trade-offs need to be made.

The outside of the circle represents the absolute limit of traction a tire can produce. If you run your tire inside the circle... You're giving up potential traction.

If you're running your car outside the circle, then you are overloading your tire and it's spinning or sliding... Losing traction.

Maximum forward acceleration is represented at the 12 o'clock part of the circle. This means you are not trying to turn the corner at all ... you want no side bite out of the tire. This represents the total amount of forward traction a tire can give you.

The 6 o'clock portion of the circle represents straight line breaking. The front wheels are perfectly straight and you're not trying to turn into the corner at all.

The 3 o'clock portion of the circle represents total side bite ... No breaking ... No acceleration.

This can be a little deceiving. You would think that if you are turning left and are off the gas and brake you are at the 3 o'clock position. But, if you're off the gas the motor is still trying to slow you down. This acts on the breaking portion of the tire equation.

This chart can represent the trade-offs in traction a car experiences as it travels around the entire track. For sake of clarity, I will use an example from corner exit to illustrate how this works.

As we exit the turn we are still trying to get side bite from the tires because we are still turning. But, we also need forward traction to get off the corner.

Let's say we plot a point on the edge of the circle halfway between maximum acceleration and maximum side bite. You will notice the tire will not produce maximum forward traction because it has to share some of that traction with side bite.

This is where the driver can make all the difference in the world. Driver can make small adjustments in his driving to allow the tire more forward traction or side bite wherever he needs to pass cars.

The straighter a driver can run his car off the corner the more forward traction he will have. The same holds true for corner entry. The straighter the driver drives into the corner, the more breaking traction the tire will produce.

Same as the case for maximum side bite. The tire won't produce as much side bite as it could if the car is trying to accelerate your break at the same time.

Now let's look at some of the ways we can put traction into the tires through car adjustments. Notice I said traction into the tires… not the car. Keep your mind centered on the tires and maintaining the contact patch when dealing with tires.

Chapter Ten

Front End Traction

We set our front ends with static measurements. Typical ones you've probably heard of or know quite a bit about are:

- Caster – the forward or rearward inclination of the kingpin access.
- Camber – the inclination of the tire (in rear view) in relation to the centerline of the car.
- Toe – the difference in measurement between the front tires from the front to the back.
- Ackerman – the change in toe as the tires are turned left to right.
- Camber gain – the amount of camber that is gained or lost as the car goes through bump and roll.
- Castor gain – the amount caster gains or loses as the car goes through bump and roll.

There are other things that go into the design of the front end like kingpin inclination, scrub radius, and caster trail, but I think that is a little beyond the scope of this book.

Front ends will scramble your brain if you let them. They can also be the source for another book entirely by themselves. I will just talk about the most relevant topics for discussion on traction in the tires here in this section and try to keep it brief enough so your not falling asleep while reading this.

The fact is that most of the adjustments that can be made are built into the car by the manufacture. These adjustments are beyond the scope of what most of what people want to get into.

There aren't too many people willing to cut up their front ends or build their own spindles.

Let's just touch on some of the things we can adjust to keep good traction in our tires.

Since front end settings affect all three areas of traction, weight transfer and wheel alignment; I'm going to break the front end settings down into sections of each adjustment and give you a little starting guide lines to follow.

Chapter Eleven

Camber

Camber and camber gain can be one of the most useful tools in the front and to create traction. Adding negative camber to the right front and positive camber to the left front will help equalize the pressure across the contact patch.

This will put more grip into the tire by utilizing the entire contact patch to it's fullest potential.

To understand how this works, take a tire out of your tire rack. Take a little air out of it so it's nice and flexible. Now, put your foot on the hub mounting plate and push outward. Noticed that a bulge forms on the inside edge of the tire where it meets the ground.

The pressure on the outside edge of the tire is more then the pressure on the inside edge.

Now, camber the tire so the top leans towards your leg and repeat the same experiment. You'll notice that the inside edge now has more pressure because the weight equalizes across the tire.

With the tire flat on the ground in tire tread surface had an equal load. As a tire flexes ... the outside loads more and the inside loads less.

With camber the inside of the tire initially has more pressure and the outside has less. When the tires flexes ... the outside gains and the inside loses ... equalizing the pressure across the surface.

Now camber is very dependent on tire structure and the type of tires you run but typically on dirt tracks camber is negative on the right front between 4° and 5°. We usually set the left front camber between 2° and 3° positive.

These are just rough beginning settings and I would use tire wear and experimentation to get closer to the optimum amount of camber you will need to run.

Here's one very interesting thing I picked up from an engineer. He was laughing at everyone running around with tire pyrometers taking tire temperatures. He said the only real accurate way to tell what a tire is actually doing with tire temperatures is dynamically as the tire is rolling around the track. Once the car slows to come into the pits the tire start to equalize across the tread surface. If you do notice a difference in temperature there are massive problems with the static or dynamic camber of the tire.

Then I got thinking on dirt the equalization has to be twice as bad. So, the only real accurate way would be through tire wear and good old fashion trial and error with good note taking.

Since we always want to think of our cars in a dynamic state, we'll look at camber gain next.

Chapter Twelve

Camber Gain

Camber gain is an essential tuning tool that needs to be understood and charted for your race car.

Camber gain by definition is actually pretty simple. It is the gain or loss of camber as a chassis goes through bump and roll.

Many people take this for granted because much of camber gain is built into the front end by manufacturers. The truth is, I believe you can make sacrifices and other front and settings to get camber gain that will put more traction into the tire. Changing control arm lengths and angles are ways you can adjust your camber gain.

When measuring camber gain remember the car rolls as well as hitting bumps in the road. A car under role will push the upper control arm to the outside further then just compressing the wheel. Depending on the length and angle of the upper control arm, camber gain on paper can actually translate into camber loss in certain situations if you're not taking chassis roll into account.

On most modern dirt cars, maintaining a positive camber in the left front can be much more frustrating than having enough camber gain in the right front.

I've seen a trend in recent years of cars not rolling off the left front like they used to. I've seen trends in left front shock absorbers that tie the left front down more than in previous years. Cars are running much flatter than they used to. This makes camber loss on the left front much less of a problem than years ago.

If you need a rule of thumb for camber gain on the right front, I would recommend 1° to 1 1/2° for every inch of shock compression. Any less than this and you will be losing camber with body roll.

You can check this pretty easily when you are doing the bump steer on your car. Get yourself a digital angle finder and take angle measurements on the plate as you compress the spindle with the jack.

Here are a couple rough rules of thumb when adjusting your camber gain on the right front.
- A shorter upper control arm will add camber gain.
- More angle on the upper control arm will also add camber gain.
- A longer lower will add camber gain, but be aware that lengthening the lower control arm will also change your track width and change many other things in your front end. It is usually not advisable to adjust the lower arm length to alter camber gain.

Adjusting camber loss on the left front is a little different story.
- Starting out with a little more static camber on the left front may do the trick. Don't get too crazy with this because with the shape of the tire it can actually loosen up the car as the tire rolls onto the center of the tire from the edge.
- A longer upper control arm will also minimize the camber loss on the left front.

Chapter Thirteen

Caster

If you not familiar with what caster is I'll give you a real quick run down before we get int how it effects the car. If you look at the spindle from the side view, the tilt of the spindle from front to back is caster. If the top of the spindle leans toward the back it is considered having positive caster. If the top of the spindle leans toward the front, this is negative caster. Most dirt cars run positive caster. Caster adds a nice stable consistent feel to your front end. Many times if your car is real darty you can either add a little caster to each side or you can add a little toe.

Adding excess caster to the front to get the caster wedging affect was popular in the 80s and 90s but isn't very popular anymore. I don't agree with the caster wedging effect and I will tell you why. If you're going through traffic and you need to turn left … then counter steer right… then counter steer again left … you only disrupt the balance of the car. In addition, under counter steer excess caster will raise the right front corner which is exactly the opposite effect we are trying to get.

If you want a rough starting point for caster I recommend 2° on the left and 4° on the right. The more caster split you put between the right and the left the more the car will want to turn into the corner on its own. I have heard some cars running a negative caster on the left and then a 2° split to the right. This is kind of normal on asphalt cars, but I have never seen or heard of this working too good on the dirt. Stability and good driver feel for the car always seems more important on dirt, so I'm not a real big fan of negative or even low amounts of caster.

Chapter Fourteen

Castor Gain

I always recommend to keep caster gain to an absolute minimum. This is for the same reason I don't agree with too much initial caster. It makes the car unpredictable and unstable. On the modifieds this is pretty difficult because of the angle of the lower control arms. If you run your upper control arms parallel to the centerline of the car ... you'll actually get a little caster loss on the right front under compression.

Most late models run their control arm pivots parallel to the centerline of the car. This provides a nice consisting caster curve and minimal caster again.

Chapter Fifteen

Toe

Toe can be a great quick fix tool if for some reason you have to make exceptions to your bump steer or ackerman setting.

The absolute most important rule of thumb I would like to express to you is that never ... ever ... let your tires toe in. I know some people do it ... I have heard stories of successful racers doing this. But, until you are racing at a masters degree level and you know more than NASCAR engineers ... I have to strongly urge you to never let your front end toe in.

This produces a very unstable effect in the car. It can produce a feeling of loose or even a tight to snap loose feeling. I've seen this firsthand and cars which don't have the ackerman right or have not set bump steer properly. Once the tires toe in they are done.

Like I said adding excess toe is like putting a Band-Aid on some of these problems. It can work but if you can optimize other settings in the car, that would be best.

On dirt late models I've worked on, where everything is optimized, a 1/4" to 3/8" is all you'll ever need.

I've heard some race car manufacturers recommend 3/4" to 1" of toe out. After I checked ackerman on one of their cars ... I found out why. Under counter steer their cars toe in. The excessive toe helps solve the ackerman problem on their cars.

Excess toe can also put a little traction in your front end by enhancing tire scrub and creating more friction between the tire and the track surface.

This is a way to put heat in your front tires if you have to run on the spec tire rule. But, you also need to think about scrub as something bad. Scrub kills speed. If you kill speed you will need excess horse power and traction to over come it.

Here is one time I would suggest adding excess toe. If you crash the car in the heat and need to get it out quickly … eyeball the toe and add a little. In this case, if the spindle is bent and you don't have time to go through everything, throw a little extra toe at it and get the car ready to race. In this case the excess toe won't kill you.

Chapter Sixteen

Bump Steer

Everyone seems to have their own ideas on what is ideal bump steer and they also have their own theories on what it does for the car. I will let you know my opinion and you can set yours to whatever you think is best.

On the right front I Bump steer cars out to about .125 to .150 in six inches of compression and the left front I bump them as close to zero as possible. Dirt cars are a little more forgiving than asphalt, but be as accurate as you want to spend the time. Most dirt guys don't take the time with this, so anything you do will most

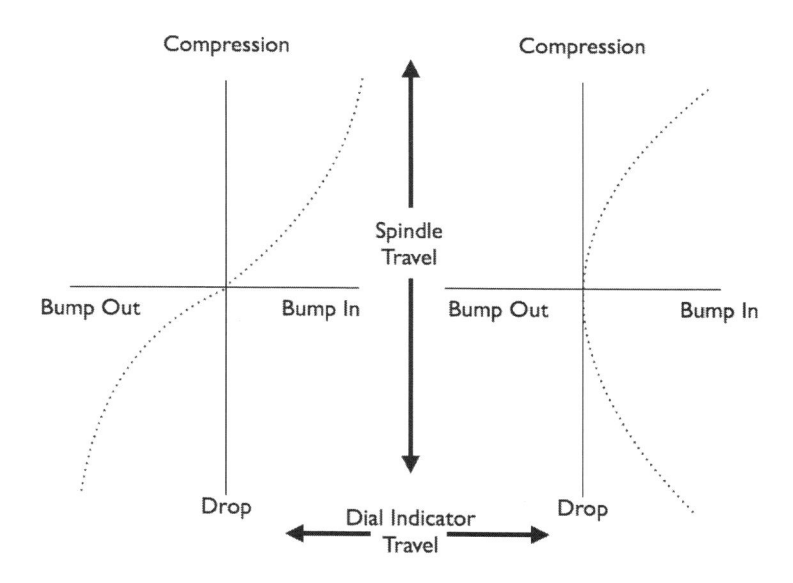

```
          Compression                    Compression
              ↑                              ↑
              |                              |
              |                              |
         Spindle
         Travel
Bump Out    |    Bump In          Bump Out   |   Bump In
              |                              |
              ↓                              ↓
           Drop         Dial Indicator     Drop
                        ←  Travel  →
```

likely be more than your competition; especially on a local weekly racing level. I would venture to guess most top level pro racers do this.

I'm going to run run through the bump steer procedure here because there are many who have never done this and whether you choose to do this or not ... it's a good skill to have. To really be at the top of your game, you really need to concentrate on the little things like this in racing.

First… I prefer to use a dual indicator bump steer set up for dirt cars. I believe the single indicator set up isn't as accurate and can give you bad results. We experience way too much camber gain to have a single indicator bump steer gauge work well.

The next thing I really need to recommend is that your stand be as sturdy as possible. If your stand flexes her wobbles as you use it you will not get good results. If your going to take the time, be as accurate as possible.

Also make sure your car is stable and doesn't wobble around. People leaning on a car as I'm bump steering it will change the measurements. Clamp your steering shaft with a couple of vice-grip pliers to make sure your wheels stay square. I've seen cars that had the steering turn while bump steer is being set. You will chase your tail trying to figure out why the adjustments you make are not making the changes they should. On a side note, figure out how to get your hubs square with the center line of the car. On the cars I build, the frame rails are parallel to the center line. I just measure off the bump steer plates to the frame rails in the front and back and get the measurements to match. Make sure this is done at ride height.

Here is a little tip that I recommend for every race team. Make yourself some right height bars for every corner of your race car.

These come in handy for different projects and you will need at least a pair for the front end to set bump steer.

Get your car at ride height and measure between the shock eyes on each corner. Get some 1/2" x 1/2" rod ends (the cheapest you can get). Then get to jam nuts for each rod end. Weld a three-quarter piece of tubing to the second jam nut on each rod end. Leave yourself enough thread so you have adjustment on the rods. Get both right and left rod ends for each rod. This will mean you don't have to disconnect them from the car to make small adjustments to ride height.

Put your ride height bars in the front end and set your bump steer gauge on the right front so everything is at zero. You can twist the face of a dial indicator so the needles line up at zero. Make sure the jack you have, to jack up the lower control arm, has enough roll so you can get 6" of compression on the right front wheel. This will be movement at the centerline of the wheel, not shock compression.

Remove the top bolt from your ride height gauge and zero the dial indicators. You might need to start the measurements on the bump steer gauge two or three inches below the "0" mark. Most commercial bump steer gauges will not bump steer a full six inches in compression.

Watching the dial indicators to see how much and which way the wheel is moving, jack up the right front slowly. Subtract the reading on the front gauge from the reading on the rear and that will give you the amount of bump steer. Pay close attention to which way the needles are spinning on the indicator. This will tell you whether you are towing in or towing out. You want to test the right front through about six inches of movement.

You can plot this out on a graph to see exactly what the curve looks like. I recommend this for new people so they can actually get a sense on what is going on with their front end. This will also help you visualize if you have a combination of problems like tie rod length and tie rod angle.

Below is an example of what your chart may look like.

Here are some rules of thumb to adjust your bump steer.

- Toe out under bump and in under drop – raise the steering arm on the spindle or drop it on the inner pivot.
- Toe in under bump and out under drop – lower the steering arm or raise the inner pivot.
- Toe out under both bump and drop – the tie rod is too long.
- Toe in under both bump and drop – the tie rod is too short.

Now, if your tire bumps out under compression more then it bumps in under drop the tie rod is probably too long as well as being too low on the spindle (too high on the inner pivot). This can be were charting the results can help identify corrections.

If your tire toes in more under bump and out under drop … The tie rod is likely to short as well is too low on the inner pivot (too high on the spindle).

I usually adjusts the right front absolute zero by adjusting the tie rod length and angle to what it needs. Then I adjust toe setting off the left front.

Adjusting bump steer on the left front is a little different. I bump the left front in both directions from zero.

I usually check the left front with more drop in mind than compression. Something like two inches of compression and four inches of drop. Follow the same basic procedure as the right front and try to get the left to bump as close to zero as possible.

It is a tedious process, but once you get this set you won't really have to check it unless you have to replace something or change you front end settings. Changes in caster and camber will affect your bump to some degree. It is very dependent on your spindle design and where your tie rod pick up point is on the spindle. Be diligent, the more you check your bump steer the easier it will get with time.

Next we're going to talk about a topic which you will have trouble adjusting, but you need to at least know what it is and how to adjust it. It's your steering ackerman.

Chapter Seventeen

Ackerman Steering

Ackerman steering is the increasing of toe as the wheels are turned left to right. You always want some positive Ackerman steering in your car. This maintains our rule of thumb that you always want your car to turn toe out.

If you are having Ackerman problems here are a couple of rules to get your things working in the right direction.

- Make sure your tie rods angle in towards the center of the car – the spindle end is further to the front end than the inner pivot side.
- Make sure your steering arm pivot of it is outside of your lower ball joint.

I like a fair amount of Ackerman especially on cars that run spec tire rules or hard tires. Ackerman can put some scrub in the front end and help build traction. Although as a side note, after a recent conversation with Brian Birkoffer, I changed the way I think about ackerman in the front.

I started taking ackerman out of the right front and reducing it to as close to zero as possible. Since you have an ⅛" to ¼" of bump out on the right front you really don't need too much ackerman. He believes too much ackerman will produce a snap loose condition under counter steer and the car will just hang with the tail out … scrubbing off traction.

On the left front, he nice thing about Ackerman is that the further you turn left the more traction the left front tire will get. This will help you turn as the car gets tight. I see ackerman on the left front anywhere between ⅜" to ¾".

As weight transfers off of the left front tire more traction is needed in that tire to help the car turn Ackerman will provide the additional traction in that tire.

There is one downfall to Ackerman. The front tires run at two different slip angles. We learned earlier that every tire has an optimum slip angle it needs to run at to get maximum traction. If you have extreme amounts of Ackerman ... You know one of your front tires will not be getting the most traction it could. One of the tires will either be under loaded or overloaded.

Steering arm length also affects where the rack is placed in the car. If you look at your car from the top down you'll notice that your tie rods angle forward to the spindles ... at least they should. Now turn the wheels left and right to make sure the tie rods never go over center and angle towards the back of the car.

This will also put a steering leverage bind in your car and the steering will not be smooth and consistent. There is a lot of force which gets loaded on the right front, especially under counter steer. Any binding will be greatly amplified under racing conditions.

If you're steering arms are short ... move your rack back in the car to make sure at full lock the tie rods are at least street perpendicular to the centerline of the car. This adjustment also will change the amount of Ackerman in the car.

Chapter Eighteen

Weight Balance and Distribution

One of the major components of race car set up is weight balance and distribution. This seems to be one of the first things we learn because 'scaling a race car' is like a buzz phrase around the racing circles. I don't want to sound cynical, but scaling is something I pay least attention to. I do think it is absolutely necessary to scale your car, but I really like to concentrate on what is going on with a car dynamically. There are too many things that can give you erroneous results when scaling a car … I try to do it as little as possible for the cars I work on.

Here are some simple notes or rules of thumb to use when scaling a car. I'm going to shoot through them real quick so we can get to the meat and potatoes, in the next few chapters, on the things we will be measuring when scaling.
- Tires will bind scales. Come up with a plan to take the bind out of scales. Rolling the car on and off the pads is good. Pulling one of the rear axles to disengage the two rear tires also works well.
- Make sure car is maintenance and ready to race. all the fuel, oil, and water is in … everything is clean and bind free.
- Scale with or without the driver, but be consistent. I prefer with the driver.
- The suspension will bind. When bouncing the car you don't need to simulate driving through a cornfield … one push in the rear and one in the front will do … do this in the same order each time.

- Make sure the steering is locked or stays in the same position. Turning the front tires will change wedge and weight distribution. This needs to be consistent.

Those are my sticking points. I'm sure you have your own or you will come up with your own over time.

Ok, let's dig into the good stuff. We'll start off with left side percentage.

Chapter Nineteen

Left Side Percentage

Let's start off with the left side percentage. I've seen numbers on winning cars anywhere between 53.5% and 54.5% on a typical four link rear suspension and as high as 56% left side on swing arm cars.

Ideally left sided percentage should be determined by the amount of weight transfer. It is very dependent on driver. Some drivers need to run a little less left side weight just to get the car to roll over. Others who turn their car a little earlier and a little more aggressively will get away with having a more left side weight.

To low of a left-sided percentage will lead to an imbalance of traction on the left side tires. To much left side and the car will resist rolling over. This also plays heavily on the center of gravity height. Drivers with higher, heavier upper body will probably need less left side than a driver who is tall from the waist down and carries a good portion of his weight below his waist.

Another aspect that affects the amount a left side percentage is the radius of the corner.

Larger radius corners allow the car more time for roll before getting to the apex and can often put less lateral load into the car. Picture it this way, on a large radius you will most likely turn the front tires less to get around the corner. This will, in turn, put less lateral weight transfer into the car. Since less is being transferred you will need to start out with less to achieve the same weight balance.

On tighter corners where there is less time from the beginning of corner entry to the apex will turn the front tires at a steeper angle and will induce more lateral weight transfer across the car and require more initial to get the same balance.

There is another aspect to consider when determining left side percentage and that is the time. I will bring it up, but I don't concentrate on it too much because I haven't found it to be as big of a factor. But, it deserves a mention.

As you turn into the corner your car will take a certain amount of time to react to the change in direction. The larger radius corner will allow the car longer to react and ultimately could end up transferring a bit more weight than the car turning a shorter radius. This is not the usual case, but I can see this happening in certain conditions and with certain drivers. I believe it is more of a driver issue than it is a radius issue. I probably shouldn't have even brought it up, but there it is.
Next item to tackle is the rear weight percentage.

Chapter Twenty

Rear Weight Percentage

Where left side weight mostly determines side bite and lateral weight transfer. Rear weight percentage usually determines forward bite and traction balance front to rear as well as having an affect on side bite. Since we after an over all balance to the car, make sure to take rear weight into consideration when balancing the car.

I've seen rear weight percentages from 54 1/2% to his high is 58%. Rear weight percentage really is determined by track radius and driver style, same as we learned with left side weight.

Rear percentage is used to tighten the car and can add tremendous amounts of traction, but the down side is having a car too tight to be maneuverable.

On large radius corners or high-speed corners, less rear percentage is needed to balance the car. Since there is a fair amount of momentum being carried in high-speed or large radius corners the car the car does not have to accelerate as hard off the corner and needs less instant traction.

Too much rear percentage can actually over tighten the car and caused a driver to wait on the car to turn. This will slow lap times. You need the car to turn as quick as possible so the driver can get back with the throttle as fast as possible. Waiting is bad.

Notice I said as fast as possible not as hard. On dirt a throttle pedal needs to be gently squeezed to maintain traction in the car. But, wasted time coasting will hurt lap times. The driver should be either on the gas or on the brake, with little lag between the two.

There are cases where more rear percentage will be faster. Paperclip style tracks will always be places were large rear percentage cars shine. The car turns quick and need lots of traction off the corner. Another rule of thumb is the more you slow down in the middle of the corner the more rear percentage you'll need to get off the corner. Rear percentage will give that car a little shot of traction to start momentum again.

I would advise to run as little we are percentage as possible as long as the driver can get traction to accelerate the car. Since the entire car is a balancing act keep the balance in mind when adding tail weight to the car.

Another aspect to consider when adding rear percentage is the pendulum effect. When you turn your car into the corner, the rear of the car becomes a large swinging weight. The further the mass of this weight is behind the rear axle the more momentum that weight will have when it gets swinging.

Here is where lead placement and the amount of fuel you will run will make a big difference. If you run your car at 55% tail weight (measured statically on the scales) and you have a full load of fuel in the cell behind the axle. Your car will handle completely different than a similar car with 55% tail and only a partial fuel load. The difference is swing weight trying to break the tires loose going into the corner.

Think about this once ... a heavy pendulum needs more energy to get it swinging and more energy to stop it swinging than the same version just lighter. So, a car with a full fuel cell will resist initial turn it more (be tighter) and want to keep sliding once it is turning (be looser), than a car with less fuel.

If you need to add rear percentage, the best place to do it is bolted to the frame above the axle or ahead of the axle.

Driver ultimately plays the largest role in what the car needs for rear percentage. Some drivers just run better with that larger amount of traction in the rear tires. These drivers tend to be less momentum drivers and more paperclips style drivers. Some drivers are a little more momentum friendly and car run right with everyone else just with less rear percentage.

Next we are going to look at the mysterious world of wedge and how it effects your car.

Chapter Twenty-one

Wedge

Wedge is the imbalance a weight between the rear tires. And most cases the left rear will weigh more than the right rear. The amount depends on many factors.
- Track configuration
- Driving style
- Suspension configuration

Wedge is used to balance the amount of grip between the rear tires dynamically. Weight will transfer left to right and front to rear in a left hand turn under deceleration. The key is to balance the weight between the rear tires for maximum grip when and where you need it. The left rear tire loses the most weight dynamically when turning into the corner. So, it is usually the heaviest static tire on the car. Coincidently the right front corner of the is usually the biggest gainer, so it usually starts out the lightest on the car statically.

Dynamic wedge can be a tricky thing to try and figure out. First as we saw earlier kinematic loading through bar angles can create a large amount of traction on acceleration. So, many times we don't need as much static wedge to get off the corner. It usually just leads to being to tight in the center. Incidentally, if your car is to tight in the center, taking some wedge out will many times cure the problem. Just make sure you don't get to tight on entry.

Anyhow, back to wedge.

We also have dynamic spring loading to consider which can be tuned to either wedge or un-wedge the car as body rolls. If the birdcage wraps against the spring as the car rolls to it's dynamic position, the load on the spring changes and the balance around the car will change.

Here is an example of what happens on the 4-link car in regards to the dynamic wedging through different spring and radius rod locations.

Almost all 4-links cars since 1997 - 1998 run with the left rear spring behind the axle tube connected to the floating birdcage. On the right rear the spring is typically mounted ahead of the axle tube on the floating birdcage.

As the left rear raises and the right rear compresses underbody roll, the birdcage will spin counter clockwise on the right rear and counter clockwise on the left rear. This will dynamically hold wedge on the left rear; and, to a much lesser extent, take wedge out of the right rear. These two actions don't cancel each other at all because typically the wedge retained in the rotation of the left rear is greater than the de-wedging of the right rear. The left rear will rotate into the spring more than the right rear.

This indexing is typically why 4-link dirt cars can run less static wedge than other types of rear suspensions; or even 4-link cars before everyone began putting the spring behind the axle tube on left rear. Dirt cars really took off when everyone began putting the spring behind on the left rear. This held wedge and traction in the left rear longer. Since wedge will make the car tight under acceleration, holding with spring indexing will keep the car tight.

We used to run wedge numbers in the 100 – 250 range. Now wedge numbers are in the 40 – 120 range.

There are several factors that contribute to the amount of spring indexing; or loading in or out of the spring as the birdcage spins.

- The distance rearward between the centerline of the axle tube and the lower spring mount.
- The distance below the centerline of the axle tube to the spring mount.
- The radius rode angles from birdcage to frame.
- The distance from radius rod mounts to centerline of axle tube on the birdcage.

Now let's look at how this whole wedge system works on the back of a four link race car.

Since, wedge is a term used to describe how the springs are loaded to distribute weight. Depending on the spring rate and the amount of wedge, the left rear spring may come off of its perch at full drop in the left rear.

If the spring is off the perch we consider this releasing the wedge. Take the bar angles out of the equation. Take your car at the attitude it runs at entering the corner and set it on scales. If the spring is still on the perch on the left rear we consider this "holding wedge".

Since wedge is technically "0" when both rear tires weigh the same, the wedge is typically long gone by the time the left rear spring is unloaded. We just use this as a term when we are describing the left rear spring load dynamically. When the spring unloads, the weight is going somewhere. Usually it's a combination between the right front and the right rear. The left front will typically see the least amount of change.

When the car is at full drop on the left rear and the right rear is compressed to a height the car is running at … measure the length of the left rear spring.

Let's say it's a 200 lb. x 14 inch spring and dynamically it's running so it's compressed to 13 1/2 inches. This means (the way I talk about it) it's holding 100 pounds of wedge. 1/2 x 200 pounds = 100 pounds.

This means that the left rear spring has 100 pounds of weight that can be redistributed to the other three corners of the car ... preferably to the right rear to aid in side bite.

The left rear will get most of its traction through bar angle. Although, having that spring with just a little tension will keep a little tension on the bars and a little tightness in the car.

It won't be to necessary to hold a bunch of wedge on that left rear. What I recommend is to adjust the left rear spring so it just touches the spring cup at full drop or just the slightest amount of preload.

This will do two things.

- It will ensure that most wedge is released and redistributed to the other three corners of the car to get a better balance of traction.
- Any amount of drop, the car will hit the spring, and not loose spring tension, bar angle, body angle on the exit of the corner where you may need all three. This will help keep the tire contact patch loaded and keep as much traction in the tire as possible.

Now, how much the left rear spring unloads also has to do with how much drop the left rear has. Everyone has their own way of determining how much wheel drop you're going to need, but one thing I do know for sure is that the left rear needs to be limited before the birdcage cams over center.

If you change radius rod angles on the left rear the amount of drop will fluctuate.

Think about the rear end actually hanging from the left rear top rod. The further up the chassis the top rod, the less the left tire will hang.

I use an adjustable chain with a clamp collar on the axle tube next to the birdcage to accomplish this. It can be made adjustable with a couple of jam nuts and a piece of threaded rod or any other system you can devise. But, make sure it is adjustable and strong enough to withstand the continual beating it will take.

If you don't adjust wedge through the left rear, how is wedge adjusted? Wedge can usually be adjusted through the other three corners of the car. You will probably need to do this simultaneously as you set ride height and the chassis angles of the car.

Lets look at some tell tail signs which will help lead you to an ideal amount of wedge for your car.

Typically, if you pick up the throttle and the car pushes or doesn't want to turn around the rest of the corner, it has too much wedge. Some how you will need to take drive out of the left rear either through bar angle or unloading the spring. If you left rear spring is already unloaded or almost all the way unloaded, I would look at bar angle.

If you lack side bite on corner entry, jack up your car and measure the left rear spring. If there is a ton of preload on that spring, that is the first place I would look to get more side bite.

You can keep playing around with wedge numbers until you get a car which will traction up nice, but yet still turn around the radius of the corner.

Next we'll look at how spring loading and wedge works on the right rear corner of the car.

Chapter Twenty-two

Right Rear De-Wedging

I look at the right rear spring indexing a little differently. I believe the right rear indexing only leads to a stiffer spring on that corner of the car. The more the birdcage rotates against the bottom of the spring as the top of the shock rolls onto the spring the stiffer that corner becomes. The rate of the spring will only accept the load you put onto it, so the more indexing you put into the right rear corner, the less body roll that corner will have.

Remember, body roll affects wheelbase change and wheel alignment. Since one small change can affect so much keep the right rear corners job as simple as possible.

I like to try and balance the right rear indexing so it doesn't overload the stiffness of that corner and prevent body roll. There is a fine line on the right rear between load on the spring and drive through the trailing arms.

I like to keep the right rear as consistent as possible. All the wedging, de-wedging, wheelbase and everything else guys try to change can all be handled on the left rear. There is no need to complicate the process by doing it with the right we're also. The only purpose of the right rear trailing arms and the right rear spring is to plant the tire and give it traction.

Period!

I hear all the time how moving right rear rods affect the wheelbase. It does, but don't move the rods to affect the wheelbase. Move the rods to put and keep traction on the right rear tire. Don't make the right rear more complicated than it needs to be.

It seems a pretty common rule of thumb is to have a 15 inch split between your mounting points on the frame for the right rear trailing arms. Many people recommend moving both arms the same to keep that 15 inch split. Remember that moving the arms not only changes the wedging and de-wedging of the spring it also affects the push point that we discussed earlier. A combination of the angle of the push point and the load through the spring is what gives and holds traction on that tire.

Next we'll look at how the right front spring affects the right rear.

Chapter Twenty-three

Soft Right Front Spring

Dynamic wedge can also greatly be affected through the front springs. A common tuning tool is to drop the dynamic wedge out of the car through softening the right from spring.

Let me explain how this works because I see so much confusion surrounding the right front spring.

First of all softening the right front corner does not put more weight on that corner. Technically softening the right front spring keeps the left front corner loaded more.

The stiffer the spring… The faster and more weight will transfer to it. In deceleration weight transfer the left front is stiffer than the right, so transfer from back to front will hit the left front corner more and harder. On lateral transfer from side to side, since the right front is softer the right rear will accept more weight than it did before and add side bite and balance traction.

What you really need to think about is how body roll and spring changes affect the attitude of the car. And, how spring placements can affect the weight loading of the car.

Our cars are welded together in one mass that connects the front and rear. It seems that each car manufacturer differ on spring stiffness and which end of the car has more effect on spring changes.

For instance, some cars really respond to small changes in the right front spring rate and other cars need a larger spring rate change before the driver can feel the difference. If very small changes in spring rate can be felt, the right front is holding up the right rear. As you soften the right front, it is effecting the right rear more.

If you make a change to the right front spring rate and the car just seems to fall on to the right front corner and not affect the right rear, the rear of the car is more likely holding up the front either through the ell center or right rear spring.

Another tell tale sign when the front is holding up the rear is when you soften the right front and you see the body roll more onto the right side instead of onto the right front.

Let's take a look at the spring placement laterally and how body roll affects this.

Look at your car from the back and judge where your right front and right rear springs are in the car in relationship to each other.

Let's say you're right rear spring is roughly 3 inches further inboard from your right front. As your car rolls onto the right side you will notice that the right front will need to compress more for the same amount of body roll on the right rear.

Now, let's see your right rear spring is tucked inside the right rear tire the same amount as you're right front spring is tucked inside of your right front tire. We will still have an offset in the springs front to rear roughly the amount the right rear tire is tucked inside the right front tire. We looked at this example in the wheel alignment section.

Okay... If we run our suspension through the same example as we did in the wheel alignment section, where we compress the right side and jack up the left rear. The right side tires will begin to get closer to lining up which means our springs will be closer to lining up.

Does it rally matter if the right side springs line up? Somewhat. I believe, the further outboard your spring placement is the more stable you setup will be and smaller changes to the right front will affect the right rear quicker. Remember, weight transfer not only depends on "how much", but "how fast". Smaller changes in body roll will have a larger change in the compression of the spring. If the spring has to be compressed to be able to hold traction in the tire, the more its compressed, the more forgiving it will be over bumps in the track.

There is an old rule of thumb which states:
- The further inboard your springs are ... the more instant traction you car will have, but it will run out down the straight sooner.
- The further outboard your springs are ... the less instant traction, but it will run out less down the straight.

Remember too, we have a shock attached to that spring. A shock further out will tend to run at a different frequency or speed range than a shock further inboard. But this is a topic to be discussed in another book.

Always look beyond the rules of thumb to setting up your car. Study how your race car works and think about your car in a dynamic state instead of your car set up in the shop and you will be able to solve any handling problem that will arise. Come up with your own ideas, model them in the shop, and test them at the track.

Let me give you a real world example of all three of our elements coming together and how to solve some simple handling problems using this method.

Chapter Twenty-four

Track Example

Okay, we go to the track with our car and our best guess set up. We take the car out for hot laps and it runs good. Nice and neutral steering with good traction.

Now is the track begins to dry out, where do we go from here?

Well let's think about everything we learned.

We want to maintain a balance set up, but we will want more traction. We won't be transferring as much weight because the lateral "G" load will be less. With the less weight transfer we will not roll the car as much. The car also won't roll steer as much. The right side tires won't align themselves the same as our neutral car in the mud. There will be less traction because of the dry track and the right rear spring will not be loaded as much. This will take side bite out. Basically the entire balance of our setup will be gone.

We can soften the right front spring which will put roll back into the front. If the front is holding up the entire car, it will put roll into the entire right side. This will bring our wheel alignment back to balance on the right side as we had when the car was in the mud. It will also load the right rear tire to put side bite and ultimately more traction back into the car. Remember traction is a combination of having both tires loaded. if the right rear isn't loaded up to its highest potential, you will be giving up traction.

The wheel alignment as well as the rebalancing of traction will bring the car back to a neutral state.

We can also alter the panhard bar position in the rear to affect body roll and increase kinematic weight transfer in the back. If our car is in balance we will need to adjust both ends of the car to get it to roll evenly onto the right side. Remember the panhard bar pinion position and the frame position really do two different things. If you simply want to put roll back into the car, lower both ends of the panhard bar evenly. Small changes of ½" will be sufficiently noticeable to most drivers.

If you just want a little more traction to the right rear and the car is rolling to the correct position, add more kinematic stick to the car. I would recommend lowering just the j-bar on the pinion instead of both ends. Pinion adjustments are very sensitive, so 1/2" increments will make a big difference.

I normally reserve the pinion panhard adjustment to just put traction in the car. But if your right rear is lacking traction this will be a good adjustment. There will be some experimentation to see what will work best for your car.

Ok, let's say the track slowed down quite a bit and the line entering the corner changed. Now to beat everyone into the corner you will have to brake hard and dive to the bottom under other cars. You will still need side bite and traction off the corner and the side bite will need to build more rapidly because of the quick dive into the corner as opposed to the slow large radius turn as before.

One thing which I have found to help people who need to do this is to increase the spring rate on the left front corner. I've personally used spring rates as high as 600 lb/in and have heard of Rockets using as high as 800 lb/in. Some really smart racers recommend against this, but this is one thing I've really seen which can help racers who like to drive deep into the corners and brake hard ... and they win a lot of races like this.

Another thing I really need to mention is strategy. Basically if everyone is on the cushion you most likely not going to pass them if your running in the same line ... unless you want to rough them up a bit. Keep on how the track is shaping up and get to know your competitors and know where they like to race. Then tune your car to pass them where they aren't going to be.

If the center of the track is slick and the cushion is big; you might not want to drop the panhard bar or soften the right front spring if you are going to have to pass cars on the cushion. If everyone is going to be on the cushion you may have to tighten your car up more and try to pass people through the slick and maybe catch some traction somewhere on the way out of the corner.

Let's say you are running a track where you turn your car early and drive your car straight through the slick because there is no traction to be had to get off the corner. Lowering the left rear bottom rod will:
- Hold wedge and give you left rear tire a little more traction by keeping spring tension on the axle.
- Take some roll steer out and help the car run straighter off the corner. (the drawback is not allowing the car to turn in as easy as when the rod is up.

Raising the left rear top rod will:
- Give a little more instance traction with the consequence of loosing a little traction further down the straight.
- Reduce the amount of drop the left rear axle will have and ultimately take some drag and aerodynamic wedge out of the body.
- This will also hold some wedge the same way as lowering the bottom rad and help you get off the corner.

Just remember there are give and takes to everything and although some tweaks may give you more traction at some points of the track, it may ultimately hurt lap times by slowing you at other parts of the track.

The ultimate goal is to have the fastest car that you can drive where you need to pass cars. Which is why a balanced setup, with a very maneuverable car is more desirable than a car that is way too tight, but has some traction. Usually you can drive a well balanced car in ways to get your lap times down even if you don't have tons of traction. Don't get me wrong, I'm not saying traction is bad, I'm saying don't sacrifice drive ability just to get a little traction off the corner.

Chapter Twenty-Five

Some Swing Arm Notes

I've decided to include some notes on swing arm suspension because I still see them being used quite a bit in the modified class. I still think they are a good suspension choice for numerous reasons, one of which because they are very forgiving and a little more consistent. The window for a missed setup is much less with a swing arm as opposed to a typical four link we see so much today.

Swing arm suspensions can be a lot different in some aspects and very similar in others. The good news is the physics are all the same. The key is understanding the bigger picture so you can make adjustments to how you would drive it and tune it to get the most speed.

Motion Ratio

The most glaring difference is that the rear springs are mounted on the lower locating linkages instead of the birdcages. Where the springs are placed on the arms is critical to the spring race of the car.

As a rear axle moves, the springs will want to move at different rates. Since springs are measured by the number of pounds it takes to deflect the spring per inch, a spring will need to be stiffer than if it were connected to the axle or the birdcage.

This is actually the case in the front suspension also but there it is rarely talked about because most lower control arms mount the springs in the same location. On modifieds spring rates are talked about in relation to the type of suspension they run on. We talk about whether it is a Chevelle suspension or a metric suspension but don't concentrate on actual motion ratio.

There is also motion ratio present on a four link rear suspension. Just like the front it is often overlooked. There is a huge tuning advantage to adjusting the motion ratio on a four link rear suspension but is beyond the scope of this section. We're just going to stick with the swing arm suspension.

There is also a motion ratio that needs to be figured from the center of the tire to where the spring is mounted laterally on the axle. This is rarely taken into consideration as far as mathematically figuring out the spring rate. But, what we do pay attention to is what we call springs stance or how far apart the springs are in the rear the car. The closer together the springs stance the stiffer the springs will need to be or the further apart the spring stance the softer the springs can be.

Let's say a spring is mounted on the axle and for sake of argument it's spring rate is 100% of the axle or tire movement. If it is a 200 pound spring the car feels as if it is a 200 pound spring.

If the spring is mounted half the distance between the axle and the chassis mounting point it has a 50% motion ratio. This means that a 400 pounds spring would need to be used to make the car feel the same as if it had a 200 pound spring on the axle. Most swing arm cars operate on roughly a 30% motion ratio which means it is located roughly ⅓ the distance between the axle and the chassis mount; closer to the axle.

If we have a 200 pound spring connected to the axle and we wanted to mount it to the swing arm one third of the distance between the axle and the frame, we will multiply the 200 pounds spring by 33% or .33. This tells us we will need to increase the spring rate by 66 pounds to support the car the same way as if it were attached to the axle.

Now don't start whipping out your calculator to figure out what spring rates you need. Figuring spring rates in the back your car is not always as simple as just doing the math. First there is quite a bit more happening in the rear suspension of a swing arm car that affects spring rates than just the placement on the lower swing arm.

Since the spring bottom is mounted on a suspension linkage, it moves in the same direction as the chassis in bump and in that roll. When the chassis rolls off the left rear tire, the bottom of the spring is moving as well as the top. This means the motion ratio is quite a bit different then if one is static and the other is moving. This makes the left rear feel much softer than the spring rate supporting it.

The same holds true on the right rear. As the car rolls onto the right rear corner it will need a certain amount of spring to support the body roll. As the spring top is moving down the spring bottom is also moving down, just not as much. The right rear will tend to feel much softer than the calculated spring rate because the spring is not being compressed as much as the chassis is moving.

Here's another aspect of motion racial to consider. The chassis is moving down onto the spring in a linear manner but, the swing arm is moving in a radius. This effect is slight but with radical link angles and short arms this could become a factor.

Here's a third aspect to consider about a swing arm motion ratio. The swing arm birdcage will still index the same as a four link birdcage just not as much. As the birdcage indexes the placement of the lower arm pivot will raise in relation to the rear axle height. This will raise the spring bottom on the axle side as it is being lowered on the frame side.

This gets way too complicated for the average racer to figure out and really not worth the effort involved. So I'm going to give you some rules of thumb to get you pointed in the right direction. Then we can talk about how to tune the spring rates in the spring section.

On the left rear of the dirt late-model or dirt modified swing arm car I wouldst start with a 350 spring mounted about 30% and on the right rear I would start at 400 palm spring at around 30%.

Many swing arm people mount the right rear spring on the front of the birdcage. This will create a stiffer right rear suspension with a much softer spring. Typical spring rates for the Z-link, as we call it, are anywhere between 200 and 225 pounds per inch. This is very similar to a typical four link right rear. The spring will ask you be compressed more than the chassis compresses on to that corner of the car. The increase is because of a different motion ratio caused by the indexing of the right rear birdcage.

Now that we have a fairly good understanding on spring rate for the swing arm let's look at push point or the instant center thrust angles swing arm suspension produces.

Instant Center

The instant center for a swing arm suspension operates in the same manner as a four link. The only difference is that since the trailing arms mom on the opposite side of the axle the change to axel movement is a little different.

First, I like to see the left rear lower swing arm mounted at about axle height off the ground or 1 inch below. The rear rod is mounted level or 1 inch below level. The idea is to try and keep the push point either above the axle in the front or below the axle in the rear. Just like four link we want to keep the thrust angle pushing up on the chassis were down on the left where suspension.

Adding thrust angle by lowering the push point in the rear will add more instant kinematic traction to the rear. The trade off to adding more kinematic traction is that it usually runs out on the straight. You really need to take into consideration where you need the traction and how the track changes throughout the night. It is also very dependent on your driving style.

Some people thrive on instant traction. So, lowering the push point on the left rear gives them a little more instant traction when the track slicks off. This is usually accomplished by lowering the left rear top rod. I've seen angles as much as 30 to 35 degrees down hill on the top rod.

Others that have run a lot of angle on that left rear tire rod thought it made their car to tighten the middle of the corner. The guys who like to pick up the gas early usually don't like lowering the top rod.

Now the right rear is different than the left. The goal is to have the right rear push point even or slightly below the centerline of the axle. The lower the push point on the right rear, the more thrust of the axle will hold up right where corner of the car. Since a lot of the side bite in the swing arm car comes from the elastic weight transfer on the right rear tire, you don't want to hold the car off of that spring. There is very little if any Kinematic site bite to load the right rear on a Z link or swing arm with a long panhard bar. A common adjustment is to raise right where Rod top rod to level.

Normally the right rear bottom rod runs uphill both 15 to 20°. With the top rod leading downhill about the same. This puts the push point well below the centerline of the axle and adds a lot of support on the gas. It also indexes into the right rear spring and makes that corner of the car stiffer. Adding even more support.

A common tuning tool when traction leaves the track is to take angle out of the right rear rods. This will take out Kinematic support as well as indexing out of the spring. This makes the right rear corner softer both and on and off the gas. This will give the car side bite in and off the corner by adding roll to the right rear that will normally diminishes as the track slows down.

The nice thing about a Z linked right rear as opposed to swing arm is the ability to adjust right rear lower rod angle without having to adjust the spring for ride height. It also keeps the push point fairly consistent in height when the right rear rods start parallel to the ground and each other.

Pull Rods

I have found no better option for consistent forward traction than the pull rod. So basically, a pull rod sounds like a no-brainer for any driver that is ever had to negotiate a slick track. Although, they do have some drawbacks. I believe the drawbacks are the reason they have fallen out of favor in the dirt late-model world.

There is a trade-off that takes place on just about everything that goes into it race car. Trends go in and out of favor when people find certain ways to run their cars and win races. Sometimes the reason they win races has nothing to do with how they run their cars. But, if they win most people will follow what they do.

    Let's look at the positives and negatives for a pull rod and you can determine whether you want to try one for yourself.

    The pull rod is a torque absorption device that connects to the top of the rear end. It serves the same purpose as a torque arm and fifth coil set up.

    The most positive aspect of using a pull rod is the instantaneous traction it supplies. The biggest and most glaring problem I see with pull rods is the pinion angle change it goes through under action. This problem becomes increasingly problematic when using a J bar instead of a long panhard bar.

    When a j-bar sees a lot of pinion angle change, it loses angle and moves to rear end to the right. This can cause the car to break traction further down the street. There is a little benefit to the pinion angle change but with a pull rod it can get excessive.

    Another negative aspect of a pull rod can be that it will hold up the car from rolling smoothly onto the right rear. This is because the frame will want stand on the pull rod and not allow the car to roll smoothly. Remember that in order to achieve good side bite we need the car to roll smoothly onto the right rear tire.

There are some solutions to some of these problems that make a pull ride very attractive. First is using your pull rod with the long panhard bar similar to the ones run on asphalt cars. These panhard bars connect to the rear end on the left axle tube and the frame on the right, usually around the right rear spring.

The frame side of the panhard bar is usually mounted 1 to 1 ½ inches higher than the left axle tube side.

What this actually does is creates a leverage to pull the car down onto the right rear spring and un-wedge the car to give it side bite. Look at your car from the back. Imagine pushing your right rear tire left as you pull the rest of the car right. This represents what happens to your car is it enters the corner. The car will want to level the two points of your panhard bar which in effect will pull down on the right rear spring and take wedge out. This will have a multiplying effect and side bite will increase the more the car is pulled into the right rear spring.

Now, there are limits to everything and if one is good two is not always better. If you raise the panhard on the frame to high the roll center will be too high to allow the car to roll over.

I usually keep the split under 2 inches. But, always keep some split between the two points. If you mount the frames side of the panhard bar below the axle pivot the force will actually want to pick up and fight the natural roll of the car.

Wheel Alignment

Another drawback to the swing arm suspension is the lack of kinematic weight transfer in the rear of the car. All of the weight transfer in the rear will be from elastic loading through the spring on the right rear. Because of this it is important to keep the right rear left of the centerline of the right front tire.

On our example with the four link, the right side tires came close to lining up in a dynamic state. This isn't the case with a swing arm. To get enough side bite, move the right rear tire inside the right front.

When using the long right side panhard which are common on swing arm cars, there isn't as much lateral movement to the rear end. Another huge difference is the lack of roll steer. This all means rear end stays pretty consistent as you view it on the garage floor.

Swing arm cars are noted for their traction. These cars are best set up and driven to use the traction to an advantage. Swing arm cars don't run as well as four-link cars on long radius corners. Because of this we don't normally set these cars up the same as four-link cars.

Typical swing arm starting percentages are 55 to 56 percent rear and 55 to 56 percent left. Left side percentage can be cut under 55 percent if you are running on a long radius corner and need the car to have as little more side bite. This is usually the case when you need to drive the car into the corner on the gas. The trick here when decreasing left side is to keep the bar angle out of the right rear and allow the car to slowly transition on to the right rear as you drive into the corner.

As far as wedge for a swing arm, I've seen wedge numbers all over the board. I've heard of people running wedge as little as 180 to as high as 300 pounds. Lower wedge numbers will probably work better for larger radius corners. On tighter paperclip type corners where you want to car to turn quickly off the gas and traction back quickly, larger wedge numbers may work better.

Remember that wedge will work in and off the corner. The higher the wedge number the less side bite it will have off the gas but the tighter it will be on the gas. This is why it works well with a paperclip style track. Once you back off the gas the car will rotate quickly and then you can accelerate quickly off the corner. The amount of time off the gas to allow the car to rotate will be diminished.

Remember you will need to experiment to find the best combination that will work for you. There are no hard rules to setting up your car. The best tools you can use is your understanding of a car to get the most out of it.

Keep experimenting and keep trying things until you hit on something which really works good for you. Then create this as you base line setup. The experiment from there to keep finding something better.

Keep good notes and if you get off base a little, you can always go back and start again from the point your car really worked good.

Chapter Twenty-Six

Conclusion

Well, we could actually go on just about forever on the topic of race car set up and how to go fast, but I really need to put a cap on it somewhere.

I really have to acknowledge some references that I used to build this book. These are not copied or plagiarized, but the concepts came from them and they deserve a plug.

The bump steer chart I drew up as an example came from Tony Woodward's steering catalog. I've been using this for years, so I decided to add it in here. Also if you have a chance to pick up one of his catalogs, do so. It is full of useful steering diagrams and information.

Second I would like to thank Claude Rouelle. I took his three day intensive workshop back in, I believe 2005. It was by far the best and most comprehensive course about race car dynamics I've taken. It is very technical and not slanted toward the dirt racing market, but I believe all race cars are created equal … at least in principle.

Third I would like to reference some good books to continue your education. Some of these aren't cheap, but these are what I reference when I run into a problem I'm really stumped on.

Any of the Milliken books on vehicle dynamics

The Rowley Race Car Engineering book. This book coincides with the use of the Bill Mitchell geometry software … also a great investment to figure out how your car works. I have it and love it.

If you really liked this book and you would like to give me a little feedback, you can visit my blog and leave your comments in the comment form in the 'contact' section. My blog is located at:

http://hogantechnologies.com

You can also sign up for my mailing list. From time to time I give away special freebies to only my list subscribers.

For instance, this very book you're reading I gave away to my list for free. If you would like to be apart of my Facebook community search for Hogan Technologies on Facebook and hit the 'Like' button. This will keep you up to date on changing technologies or any new ideas which seem to be coming around in the world of racing technology.

Thanks

Be FAST!

Kevin Katzenberg

Made in the USA
San Bernardino, CA
12 December 2013